SpringerBriefs in Physiology

For further volumes:
http://www.springer.com/series/10229

Linda E. May

Physiology of Prenatal Exercise and Fetal Development

 Springer

Linda E. May
Kansas City University of Medicine
  and Biosciences
1750 Independence Avenue
Kansas City
MO 64106-1453
USA

ISSN 2192-9866   e-ISSN 2192-9874
ISBN 978-1-4614-3407-8   e-ISBN 978-1-4614-3408-5
DOI 10.1007/978-1-4614-3408-5
Springer New York Heidelberg Dordrecht London

Library of Congress Control Number: 2012933847

Printed on acid-free paper

Springer is part of Springer Science+Business Media (www.springer.com)

Preface

In the quest to provide children with optimal health, maternal exercise while in utero is sometimes not considered. Concerns have been raised that the physiologic effects of maternal exercise on thermal equilibrium, placental bed blood flow, and delivery of oxygen and nutrients to the fetus may compromise the growth and development of the fetus. However, collective findings have supported that acute cardiovascular, hormonal, nutritional, thermoregulatory, and biomechanical responses of exercise are not teratogenic, and do not compromise the pregnancy. Moreover, results of further studies revealed that physical exercise provides benefits for expectant mothers, such as an improved sense of well-being, diminished discomfort and pain associated with pregnancy, and maintenance of maternal cardiovascular health. Exercise during pregnancy also maintains and/or improves maternal fitness and physical capacity, significantly decreases risk of developing preeclampsia, hypertension, gestational diabetes mellitus, decreases weight gain, and improves attitude and mental state (Collings, Curet et al. 1983; Dye, Knox et al. 1997; Clapp, Kim et al. 2000; Pivarnik 2006). Placental adaptations have been discovered that indicate an increase in blood flow and nourishment to the fetus. Mothers who are well-conditioned and continue their exercise regimen throughout gestation show no increase in lack of conception, abortion, congenital abnormalities, or preterm labor, and have normal growth and development during the first year of life (Clapp 1991, 1998).

In light of the information that maternal exercise is not harmful to the fetus, while also being beneficial to mother and the placenta, our research has begun to examine pregnancy outcomes as they are related to exercise benefits of the fetus/infant. Throughout the Chap. 1 we will describe current findings of how maternal exercise throughout gestation influences fetal development of key organ systems. The Chap. 2 will explain how these effects influence the offspring during labor and delivery. This chapter will also examine the relationship between maternal activity level and perinatal cardiac autonomic control of the offspring. Chapter 3 will describe the longitudinal effect of maternal exercise on postnatal growth. These chapters will encompass the relationship between maternal activity level and fetal, birth, and neonatal effects.

Some scientists have collected data on offspring exposed to maternal exercise in utero. Although methods and models have differed, these findings in animal and human models are helpful as we further investigate this topic. To this point, all of this research has not been summarized. Therefore, the Chap. 4 will focus and summarize studies related to physical/health measurements of offspring exposed to maternal exercise throughout gestation.

The next part of the book will describe the most effective way to address barriers to women exercising during pregnancy, and maintaining a physical fitness program. Additionally, the most effective and safe exercises during pregnancy will be put forth (Smith 2008). Previous guidelines have become less conservative based on research studies. In order to optimize safety, while maximizing benefits, current guidelines based on federal, American College of Obstetrics and Gynecology (ACOG), and the Canadian Society of Exercise Physiology (CSEP) recommendations will be discussed in the Chap. 5. Various aspects of exercise which have been studied include frequency, time, intensity, type of exercise, and environment. These exercise factors will be described in relation to pregnancy outcomes.

By the end of the book, readers should better understand the newest research findings on how exercise influences the fetus in utero and beyond, the most effective ways to maintain an exercise protocol for pregnant women to exercise, and the specifics of an appropriate exercise routine. This information will help researchers and scientists better understand the effects of exercise during pregnancy on offspring development.

Contents

Chapter 1
Maternal Exercise Throughout Gestation and Fetal Development

Abstract Initial research established the safety and efficacy of maternal exercise during pregnancy for mother and even the placenta. But what about baby? The next logical progression of research interest is what effect maternal exercise has on the fetus and its development. In order to investigate this further, research will need to control for factors that are known to influence activity state of the fetus, such as time of day, or recordings, as well as time since last maternal meal/snack. It is suggested that acute or chronic maternal exercise does not impact the fetal activity state. Though it appeared that acute maternal exercise may affect fetal breathing movements, many factors were not controlled. To further elucidate, follow-up studies focused on exposure to regular maternal exercise throughout gestation found no evidence of fetal distress or lack of oxygen throughout the pregnancy. The plethora of research clearly shows that the fetal heart is able to respond during acute bouts of maternal exercise. The effects of this response were further illuminated by a study of chronic exposure to maternal exercise. Fetal heart outcomes due to regular maternal exercise are similar to adult heart response to exercise training. Whether fetal motor activity is influenced by regular maternal exercise is an area that still needs further study. Lastly, the evidence of fetal nervous system development points to properly advancing gestation and possibly a more advanced trajectory. Overall, fetal development is not adversely affected from exposure to maternal exercise.

Keywords Fetal development · Fetal activity state · Fetal breathing activity · Heart rate (HR) · Fetal motor activity · Nervous system development

L. E. May, *Physiology of Prenatal Exercise and Fetal Development*,
SpringerBriefs in Physiology, DOI: 10.1007/978-1-4614-3408-5_1,
© The Author(s) 2012

Background

Research studies on exercise during pregnancy have increased exponentially from 4 publications between 1946 through 1950 to 208 publications in 2011 alone (pubmed search). Originally, researchers focused on whether exercise during pregnancy was safe for mother and fetus. Once the safety of exercise during pregnancy was established, researchers then began to study whether physical activity was beneficial for the mother, and/or the placenta. As a result, it is now known that maternal exercise benefits the maternal-placental unit (Clapp 2000, 2003, 2006; Clapp et al. 2000, 2002, 2003). Current research is aimed at how maternal exercise affects fetal development.

For researchers there is a dilemma: our study subject cannot be seen, nor heard, barely felt, and some of our best measures (i.e., birth weight) do not tell us much about health (Sontag and Huff 1938). What types of data can we collect noninvasively during the pregnancy that will provide information about the health and well-being of the fetus. Efforts have focused on investigating the effects of exercise on a few aspects of fetal development: activity state, fetal breathing activity, heart rate (HR), motor activity, and nervous system development. The influence of acute and/or chronic exposure of maternal exercise on these fetal measures will be discussed.

Fetal Activity State

Based on extrapolating data from the neonatal state, fetal activity state is a way to classify quiet (less lively) versus active (more lively) fetuses. Fetal activity state determination is based on the fetal HR pattern, body movements, and, if available, eye movements as well (Nijhuis and Ten Hof 1999). Quiet activity states, IF and 3F, are associated with absent or few HR accelerations, low amplitude body movements, and less eye (non-REM) movements (Oppenheimer and Lewinsky 1994). Active activity states, 2F and 4F, are associated with the presence of more frequent and longer lasting HR accelerations, more frequent and vigorous body movements, as well as more eye movements (Ten Hof et al. 2002).

Few studies have analyzed whether maternal exercise alters fetal activity state. Researchers have noted that acute maternal exercise does not alter the fetal activity state from that which was noted prior to the exercise session (Spinnewijn et al. 1996). Scant research has evaluated activity states of the fetus in light of chronic exposure to maternal exercise. Chronic exposure to maternal exercise does not seem to influence the pattern of presence of activity states in the womb (unpublished data). Since fetal HR and body movements are used to determine activity state, we looked for differences for these measures within each state (active and quiet). Within each fetal activity state, no differences are seen in maximum duration of HR increase, or presence and amplitude of body movements (unpublished data). As gestation progresses, motor patterns occur less often, but are more vigorous; this change is indicative of appropriate fetal maturation (DiPietro et al. 1996). There are no differences in this natural development occurrence with regard to chronic exposure to maternal exercise (unpublished data). These data suggest that chronic maternal exercise does not have an impact on fetal activity state. In order to investigate this area further, researchers will need to control for factors that are known to influence activity state of the fetus, such as time of day, or recordings, as well as time since last maternal meal/snack.

Fetal Breathing Movements

Less common than fetal heart rate (FHR), fetal breathing has been routinely measured to determine fetal well-being during pregnancy. Though the fetus does not practice breathing movements continually, these movements are essential for proper lung development. A fetus in distress, for example, will have decreased breathing movements in order to conserve oxygen supply. Although two studies reported decreased fetal breathing immediately after maternal exercise, these studies each used a small sample of women, and did not control for fetal activity state (Winn et al. 1994; Hatoum et al. 1997). One study found no association between maternal exercise and fetal breathing movements during an exercise bout (Platt et al. 1983). Marsal, Lofgren and Gennser, (Marsal et al. 1979) and Manders, Sonder, Mulder, and Visser (Manders et al. 1997) found an increase in fetal breathing movements immediately following a session of acute maternal exercise. Though these last two studies did not control for fetal activity state either, they occurred earlier in the third trimester and represent three times as many pregnant women. Most likely these differences represent differences in fetal activity state and possibly time of day.

To further elucidate these findings, other research looked at chronic maternal exercise and fetal oxygenation. Regular aerobic exercise during pregnancy increased resting maternal (and possibly fetal) plasma volume, intervillous space blood volume, cardiac output, and placental function (Clapp 2003). These physiological changes safeguard against possible acute reductions in oxygen and nutrient delivery during exercise and most likely increase oxygen and nutrient delivery to the placental site after the acute exercise session (Clapp 2003). Thus, the effect of maternal exercise on fetal oxygenation is dependent on frequency and duration of the exercise (Clapp 2003). Analysis of amniotic fluid erythropoietin, indicative of oxygenation throughout the entire pregnancy prior to labor, is not different between fetuses exposed or not to regular maternal, aerobic exercise (Clapp et al. 1995). Combined, these findings suggest the fetus is not in distress and does receive ample oxygen and nutrients.

Fetal Cardiovascular Development

One of the most easily measured aspects of fetal health is HR and its variability. Even before ultrasound, FHR and its beat-to-beat variations, called heart rate variability (HRV), have served as a basis for the clinical evaluation of fetal well-being. Since cardiac function is regulated by the autonomic nervous system (ANS), recordings of heart rate and its variability provide evidence of ANS development and regulation (Vanderlei et al. 2009). HR and HRV thus are noninvasive measures of physiological health, training, or pathology (Vanderlei et al. 2009).

Fetal HR and HRV: Acute Response

Numerous studies have evaluated the fetal heart response during acute bouts of maternal exercise. A few studies found no change in FHR during or immediately after a maternal exercise session (Marsal et al. 1979; Paolone et al. 1987; Paolone and Shangold 1987; Artal et al. 1995; Bonnin et al. 1997). Some studies found a few cases of decreased FHR immediately after a single session of maternal exercise (Jovanovic et al. 1985; Watson et al. 1991; Manders et al. 1997). In Manders et al. (1997) this bradycardia occurred at 89 and 99% of maternal maximal increase in heart rate and was accompanied by drastic attenuation in fetal HRV, along with no body and breathing movements for 20 min, while in Watson et al., the fetal HR decrease was brief and followed by increased fetal HR about 10 bpm above baseline levels (Watson et al. 1991). From Jovanovic et al. (1985) the decreased fetal HR was during exercise, but increased after exercise. Regardless of whether a woman is conditioned or not prior to conception, the majority of studies found that an acute bout of maternal aerobic exercise elicits an increased FHR response (Hauth et al. 1982; Collings et al. 1983; Clapp 1985; Collings and Curet 1985; Cooper et al. 1987; Steegers et al. 1988; van Doorn et al. 1992; Clapp et al. 1993; Webb et al. 1994; McMurray et al. 1995; O'Neill 1996; Manders et al. 1997; Brenner et al. 1999; MacPhail et al. 2000; Kennelly et al. 2002b).

Similarly, fetal HR changes from baseline in FHR were greater after a cycling than a swimming session; though the average maternal maximal HR was similar between sessions (Watson et al. 1991). Besides aerobic exercise, increased fetal HR was also seen during maternal strength conditioning sessions (Avery et al. 1999). Although the majority of findings suggested that acute maternal exercise is associated with an increase in FHR, thus far it seemed that there was no consistent trend related to acute maternal aerobic exercise and FHR. However, these differing findings are most likely due to differences in methodology, such as exercise type, maternal positioning, recording instruments, timing during gestation, and fetal activity state.

Further study findings help us to better understand whether there is a relationship between acute maternal exercise and fetal heart response. Since increased fetal HR is the most common finding during or immediately after acute exercise, what might this response mean for the baby. First of all, although fetuses demonstrate a brief elevation in heart rate with maternal exercise, researchers have noted no increase in fetal distress (Spinnewijn et al. 1996a, b; Avery et al. 1999; Kennelly et al. 2002a, b). Furthermore, a decreased incidence of abnormal fetal heart patterns was noted with maternal exercise (Spinnewijn et al. 1996a, b; Avery et al. 1999; Kennelly et al. 2002a, b). One study, taking measurements immediately after acute maternal exercise, observed that the increased fetal HR is accompanied by lower HRV (Manders et al. 1997). This study noted fetal HR increased with the increased maternal exercise level (Manders et al. 1997). This is a response typically seen in an adult during an acute bout of exercise (Bailon et al. 2003). For adults, this response is related to the level of circulating catecholamines which stimulate the heart and lead to increased HR and lower HRV. Interestingly, maternal catecholamines also increased with maternal exercise level and catecholamines are measured in the amniotic fluid (Cooper et al. 1987).

Fetal HR and HRV: Chronic Response

Taken together, the fetal heart response to acute maternal exercise is intriguing. The next logical step is to look at chronic exercise rather than just acute exercise. With chronic exercise exposure there are specific physiological cardiac responses that are typical in adults, such as lower HR and increased HRV at rest. Since pregnancy is a longitudinal process lasting about 40 weeks, it offers an interesting type of exercise training study.

It is important to keep in mind that ANS innervations with the heart occurs in late pregnancy (~ 30 weeks gestation) and continues until after birth (Renou et al. 1969). Therefore, proper heart and ANS development are indicated by decreasing fetal HR and increasing HRV with increasing gestational age (Brenner et al. 1999). Additionally, it was previously mentioned that fetuses can be in different activity states which are noted by differences in heart rate, such that quiet fetuses will have lower heart rates relative to fetuses in an active state. Since an exercise training response is associated with lower HR and increased HRV at rest, it is essential in studies during pregnancy to control for factors such as gestational age and fetal activity state.

One study compared HR and HRV in fetuses of regularly exercising women and fetuses of non-exercising women in late pregnancy. This study found lower HR and increased HRV of fetuses exposed to chronic aerobic exercise at 36 weeks gestational age (GA) relative to same age fetuses in the same fetal activity state, not exposed to maternal exercise (May 2010a, b). For comparison, fetal HR which is clinically too slow, defined as bradycardic, is <120 bpm, is coupled with lower HRV. Lower HR, but with lower HRV, is indicative of negative exposures, such as hypoxia from nicotine exposure (Mochizuki et al. 1985; Graca et al. 1991). Similarly, fetuses exposed to alcohol have lower HR, with little or no HRV (Halmesmaki and Ylikorkala 1986) and are associated with poor pregnancy outcomes. Since this lower HR is not bradycardic and is accompanied by increased HRV, it suggests that exposure to chronic aerobic maternal exercise is more like the lower HR and increased HRV associated with an adult's exercise training response.

Fetal HR and HRV: During Fetal Breathing and Non-Breathing Periods

Fetal breathing movements are an indicator of proper lung and diaphragm development as well as central nervous system development involving the phrenic and vagal nerves. Fetal breathing movements increase with gestational age as proper maturation progresses. Similarly, maturation of cardiac ANS progresses with gestation. Measurements involving fetal HR and HRV during fetal breathing movements in late pregnancy could be an indicator of fetal well-being and central nervous system development. An analysis was done at 36 wk GA to determine differences between maternal exercise exposure and fetal breathing movements on HRV. The study found significant interaction between maternal exercise and fetal breathing on HRV. These differences suggest the fetus of the exercising mother has better vagal control during breathing activity. On the whole, maternal exercise suggests improved maturation of the cardiac autonomic nervous system in fetuses exposed to maternal exercise, as well as augmented fetal, cardiac, parasympathetic control during fetal breathing movements for the exercising cohort.

Other Fetal Heart Measures

In addition, the lower HR and increased HRV are also associated with differences in cardiac time intervals. Fetuses exposed to maternal exercise tend to have increased duration of PR interval and QRS complex (unpublished data) relative to those not exposed to maternal exercise. These findings further suggest augmented cardiac maturation and/or increased muscle mass in fetuses of exercising mothers (Sharieff and Rao 2006).

The research literature in adults also shows associations between physical activity and physiological responses vary as a function of physical activity duration and intensity (LaPorte et al. 1985). Women who participated in higher intensity physical activity had fetuses with lower HR and greater overall HRV. Furthermore, increased duration of physical activity that women performed throughout the third trimester was associated with increased fetal HRV. One study (May et al. 2010a, b) found a dose-response relationship between level of maternal leisure time, physical activity and fetal, cardiac autonomic response, similar to the adult cardiovascular response to chronic exercise. Although these findings warrant confirmation from other studies, their implications are exciting considering the potential positive effects of physical activity during pregnancy on fetal cardiac ANS development.

Other cardiovascular parameters provide insight into the effects of maternal exercise on the developing fetus. Other measurements related to heart and vasculature are cardiac output and vessel resistance. Cardiac output, a measure of heart function, is calculated by HR multiplied by stroke volume. Although changes in fetal HR have been shown to change during a maternal exercise session, fetal cardiac output is stable (Chaddha et al. 2005). It has been shown that simultaneous activation of the vagal and sympathetic cardiac nerves results in greater cardiac output and lower HR (Koizumi et al. 1982). While power in defined frequency bands cannot be used as a proxy for cardiac output, the finding is intriguing and warrants further investigation. Further, fetal internal carotid artery mean velocity increased and fetal cerebral resistance index decreased suggesting vasodilation (Bonnin et al. 1997). The evidence overwhelmingly shows that maternal exercise does not have adverse fetal cardiovascular effects (Kennelly et al. 2002b).

Fetal Motor Activity

A few studies have looked at the impact of acute maternal exercise on fetal movements. Two studies showed that fetal body movements decreased during maternal exercise (Winn et al. 1994; Manders et al. 1997). Other research found fetal movements during maternal exercise; these movements are not always associated with increases in FHR not accompanied by fetal heart increases, (Platt et al. 1983; Hatoum et al. 1997). Similarly, another study noted increased fetal motor activity with maternal exercise; this observation concurs with the relationship between maternal sympathetic activity, indicated by epinephrine level, and fetal motor activity during acute exercise (Platt et al. 1983). The varying findings in these studies are most likely attributed to different exercise protocols, different gestational ages of the fetus, and other maternal factors.

There is scant research available as to how fetal movements are affected by chronic maternal exercise. One study, which looked at the effect of regular maternal exercise throughout gestation on fetal heart development, made observations of fetal motor activity at rest as well. Within each fetal activity state no differences are seen in controls and exercisers at each gestational age for high and medium amplitude body movements, and duration of body movements (unpublished data). Controlling for fetal activity state, the number and frequency of fetal body movements is similar between fetuses regardless of exposure to chronic, aerobic maternal exercise (unpublished data). Although there were no differences in fetal motor activity regardless of exposure to maternal exercise, the sample was small and the study was not designed to test for those differences. Altogether these data may suggest maternal exercise does not negatively impact the quality and quantity of fetal body movements within activity states.

Since fetal body movements are coordinations between muscle development and appropriate innervations, more coordinated movements should be seen with advancing gestation. Other research has shown that FHR and fetal movement become increasingly integrated with advancing gestation and development. Both these measurements are linked to the central nervous system development. Data have shown that fetal motor activity is influenced by maternal factors (i.e., age, BMI) as well.

Whether fetal motor activity is influenced by regular maternal exercise exposure still needs further study. A study evaluated the relationship between fetal motor activity late second and third trimester gestation and offspring neurobehavioral up to 2 years of age. Researchers concluded that fetal motor activity appears to predict temperament attributes related to stability and regulatory behaviors (DiPietro et al. 2002a, b). Another study documented an association between FHR and fetal body movement. Additionally, these reflect the development of the central nervous system, and that this development may be affected by maternal factors. (DiPietro et al. 1996). Since fetal motor activity may be associated with temperament and behavior in early childhood, this suggests the influence of maternal exercise on this measure warrants further study (DiPietro et al. 1996, 2001, 2002a).

Fetal Nervous System Development

Three measures of fetal neurobehavioral functioning are utilized to reflect newborn neurological maturation. These fetal measures are fetal HRV, fetal movement, and the coupling of fetal motor activity with fetal HRV (DiPietro et al. 2010). Since the autonomic nervous system is involved in regulation of most physiological processes, such as heart rate control, analysis of fetal HR and HRV provides evidence for fetal well-being and appropriate fetal central (i.e., autonomic) nervous system development. The measures of HR and HRV are used during pregnancy to determine overall health and appropriate nervous system development of the fetus, which can be extrapolated to relate to later development after birth. Therefore, fetal HR and HRV continue to be the most prominent and easily accessible indicator of fetal well-being (DiPietro et al. 2007). There has been little evidence to suggest that exercise regimens studied to date have adverse effects on maternal or fetal health. (Clapp and Little 1994; Brenner et al. 1999; Kennelly et al. 2002b; Gavard and Artal 2008). Fetal motor activity requires development of the nervous system and musculature (Nijhuis et al. 1984). Therefore, fetal movement is associated with activity levels in childhood and predicts child temperament (DiPietro et al. 2002a, b). Lastly, the combination of FHR with motor activity and their synchrony is associated with parasympathetic control and evidence of properly advancing gestation (DiPietro et al. 2010).

Further Research

The vast majority of information available concerning the effects of maternal exercise on the development of the fetus are related to acute changes during or immediately following maternal exercise. Studies that have examined the effect of regular, repeated exposure to maternal exercise find FHR and HRV changes, as well as neurodevelopmental and behavioral differences. Ultimately, analysis of fetal HRV during pregnancy provides a unique, noninvasive measure of fetal health or disease. These studies intimate that maternal exercise has significant positive benefits for the fetus, who may adapt in utero, and that these benefits may extend well past birth. Since there is little research concerning the long-term changes in offspring well-being in relation to the mother's exercise, further research should be conducted as to the scope of these changes in the fetus, and how they affect neonatal and childhood development. Additionally, as technology improves methods for fetal monitoring fetuses, attention should be given to how maternal exercise affects fetal behavioral state and responsivity to stimulation. This research is especially important as we begin to discover how exercise may cause alterations in fetal development that extend into adulthood. As our understanding of maternal exercise and fetal development improves, we will be able to better advise pregnant mothers as to safe and beneficial types and levels of exercise.

Future research is needed to determine if fetal exposure to exercise during pregnancy is the earliest intervention to improve child health, such as cardiovascular health, improve nervous system maturation. Considering that heart disease is a leading cause of death in the US, such research is necessary and essential.

Chapter 2
Effects of Maternal Exercise on Labor and Delivery

Abstract Despite popular belief, exercise while pregnant does not initiate labor, or preterm labor. Exercise is associated with delivering closer to the due date. Further, maternal exercise is associated with either no change, or shorter labor and delivery overall. Women who exercised had less complications during labor and delivery than those who did not exercise. More importantly, measures of the fetus during the labor and delivery process did not indicate fetal distress. Measures of fetal distress showed either no difference or improvements related to exercise exposure, such as increased apgar scores. Exercised mothers and their children had shorter hospital stays relative to nonexercisers. Overall, these findings of possibly shorter labor, less preterm labor, less complications, and decreased hospital stay all add up to decreased health-care costs. Further research is needed in this area.

Keywords Preterm · Post-term · Gestational age · Labor · Delivery · Fetal distress · Apgar scores

Finally, the time comes for the baby to arrive. Labor and delivery is one of the most physiologically challenging times in woman's life. The demands of the labor process are strenuous for mother and child. In light of this fact, it is common to monitor mother and child during this event and immediately after to ensure both have endured the stress well.

Popular belief among pregnant women and the public is that exercising will initiate labor, i.e., preterm labor. Preterm labor is defined as delivering before 37 weeks gestation. Full term is defined as 37–42 weeks gestation at delivery. Post-term delivery occurs after 42 weeks gestation. The length of gestation, or gestational age, refers to how long the fetus grows in utero. The general idea is that the longer the fetus spends in utero, more development will occur and the likelihood of survival and improved health at birth increases.

L. E. May, *Physiology of Prenatal Exercise and Fetal Development*,
SpringerBriefs in Physiology, DOI: 10.1007/978-1-4614-3408-5_2,
© The Author(s) 2012

11

Many studies have found maternal physical activity during pregnancy is not associated with the occurrence or increased occurrence of preterm labor (Curet et al. 1976; Collings et al. 1983; Jarrett and Spellacy 1983; Botkin and Driscoll 1991; Hatch et al. 1993; Clapp 1994; Florack et al. 1995; Sternfeld et al. 1995; Horns et al. 1996; Rao et al. 1998; Clapp 2000; Magann et al. 2002; Leiferman and Evenson 2003; Misra et al. 2003; Gavard and Artal 2008). Likewise, research reporting gestational age at delivery finds no difference between neonates of exercising or non-exercising women (Koemeester et al. 1995; Sternfeld et al. 1995; Horns et al. 1996). Further, studies found the risk of preterm delivery decreased related to physical activity throughout pregnancy, whereas sedentary activities, mainly TV watching, increased risk of preterm birth (Misra et al. 1998; Clapp 1991, 1998). This risk may be decreased by almost 50% after controlling for maternal body morphometry and socioeconomic status (Juhl et al. 2012; Domingues et al. 2008). Further, there is no association with a woman's activity prior to pregnancy and likelihood of preterm or post-term delivery (Gavard and Artal 2008). One study found women who exercised during pregnancy began labor significantly earlier than those who did not exercise during pregnancy, though all babies were born full term (Clapp 1990). Another study showed similar trends (p = 0.06), a tendency toward earlier (∼5 days) birth and decreased gestational age, though these values are still within the normal ranges (Jarrett and Spellacy 1983). Since the initiation of labor and delivery can be considered a communication of physiological "readiness" between mother and child, these data further support the thought of improved maturation of the exercise-exposed fetus. Thus, participation during pregnancy, not before, increases a woman's likelihood of delivering closer to her due date.

About as long as an average runners marathon finish, the entire labor and delivery process for nulliparas lasts approximately 9 h, while multiparas lasts about 7 h (Murray and Huelsmann 2009). One study suggests that exercise during pregnancy may result in shorter labor and delivery for all women (Clapp 1994). Another study found shorter labor in multiparas, but not nulliparas (Pomerance et al. 1974). On the other hand, other studies found no difference in length of labor regardless of activity during pregnancy (Botkin and Driscoll 1991; Kardel and Kase 1998). There are other factors, such as mode of delivery, which make it even more difficult to conclusively state how maternal exercise during pregnancy influences the length of labor and delivery.

Research has been done to determine if exercise during pregnancy affects the mode of delivery. Some studies have found no relationship between mode of delivery and maternal activity during gestation (Sternfeld et al. 1995; Horns et al. 1996; Kennelly et al. 2002b; Magann et al. 2002; Barakat et al. 2009b). Other research, however, has found participation in exercise while pregnant is associated with lower rates of cesarean section and less complications (Curet et al. 1976; Dale et al. 1982; Hall and Kaufmann 1987; Clapp 1990; Clapp and Capeless 1990; Hatch et al. 1993). Women who exercise during pregnancy also report lower perceived exertion during labor, than women who did not exercise (Rice and Fort 1991).

More importantly, the influence of labor and delivery on fetal outcomes can provide insight as well. Evidence of fetal distress can be determined based on clinical markers prior to and after birth. Common markers used to evaluate distress of the fetus are meconium staining, fetal heart rate patterns, fetal oxygenation, cord entanglement, and apgar scores (Clapp 1991). In fact, studies have shown that there is a significant reduction in the incidence of the clinical markers of fetal stress/distress during the labor of women who exercised (Clapp et al. 2003). Research has noted that there is neither an increase in fetal distress, nor a correlation with incidence of fetal distress and maternal exercise (Clapp and Capeless 1990; Kennelly et al. 2002b). Evidence purports no difference in fetal oxygenation during labor regardless of exposure or not to exercise throughout gestation (Clapp 1990, 1994; Clapp et al. 1995). Even walking during labor is associated with more favorable fetal oxygenation and heart rate, whereas in the classical supine and side position there seemed to be lower fetal oxygenation (Braun et al. 2004).

This is evidenced by a decreased incidence of meconium staining, abnormal fetal heart patterns, and cord entanglement (Clapp 1990, 1994; Clapp et al. 1995). Anecdotal stories of women who exercised while pregnant describe a scenario in which the health care provider noted lower fetal heart rate during labor and delivery. Based on typical paradigm this was cause for concern. However, these anecdotal stories have all had the similar outcome of a baby born in good health. These stories lead one to ponder if the lower HR (and increased HRV) seen in the fetus as a result of maternal exercise may be enough to be noted during labor and delivery. If this is indeed the case, then health care providers may need to consider maternal activity level when analyzing fetal heart rate and its variability during labor and delivery; especially as it influences the providers' decisions of course of action for mode of delivery. A uniquely designed study found that an automated quantification of fetal heart rate accelerations and variability are predictors of apgar scores and to a lesser degree fetal academia (Ayres-de-Campos and Bernardes 2004; Ayres-de-Campos et al. 2004). Considering offspring of women who exercised during pregnancy have improved fetal heart maturation and similar or increased apgar scores relative to offspring of women who did not exercise, the utilization of automated quantification may be a better predictor of pregnancy outcome. This is an area where further study is necessary for the field of obstetrics. A few researchers are beginning to study this area of fetal health during labor and delivery.

The first postnatal indicator of offspring well-being is the 1- and 5-min apgar scores. Many studies show that maternal exercise has no effect on neonatal apgar scores (Curet et al. 1976; Collings et al. 1983; Clapp and Dickstein 1984; Botkin and Driscoll 1991; Kardel and Kase 1998; Kramer 2000; Marquez-Sterling et al. 2000; Magann et al. 2002). Even obese women who exercised while pregnant delivered babies who had normal apgar scores (Lindholm et al. 2010). While two studies found improved apgar scores related to maternal exercise (Hall and Kaufmann 1987; Clapp and Capeless 1990).

The last indicator of maternal and fetal health related to the labor and delivery process is the length of hospital stay. Women who exercise while pregnant have notedly decreased length of hospitalization after delivery (Hall and Kaufmann 1987). These findings not only speak to the overall physiological healthiness of mother and child, but also influence overall health care costs. Thus, regular physical activity during pregnancy, either does not affect or is beneficial for labor and delivery outcomes and may reduce health care costs associated with this event. In summary, these data of possibly shorter labor, less preterm labor, less complications, and decreased hospital stay all add up to decreased health care costs.

Chapter 3
Fetal and Postnatal Growth and Development

Abstract The influence of maternal exercise in utero suggests there are no adverse effects on fetal development, possibly even benefits. Based on the idea of prenatal programming, what effect does exposure to maternal exercise have after birth and into childhood and adulthood. Without controlling for many factors, it is difficult to determine the effect of exercise on birth weight. At various stages of childhood and into adulthood, there are no indications of obesity in offspring exposed to exercise in utero. Birth length and height into adulthood is not adversely affected regardless of whether women exercise while pregnant or not. Little has been done in relationship to organ development. Data on rodents show no adverse effects on gross development of organs in response to maternal exercise. In humans, maternal exercise is not associated with congenital abnormalities in offspring and may even help prevent them. Generally, maternal physical activity throughout gestation does not negatively impact growth and development in utero or after birth.

Keywords Birth/body weight · Birth length · Height · Organ development

The effects of exercise on youth and adolescent growth and development have been studied from the context of: how much should be done, when should it occur, does it improve growth of muscles and bones, does it decrease fat stores, etc. However, many of these same principles cannot be directly applied to the fetus. The prenatal period represents a unique stage of anatomical and physiological development, such that perturbations to the normal in utero environment may adversely affect development.

Recent studies have indicated that exercise during pregnancy, both acute and chronic exposure, influences the fetus, in utero and after birth (DiPietro et al. 2007). Fetal maturation can be determined by growth (i.e., increased hypertrophy or hyperplasia) and development (i.e., functional differentiation). Keeping this idea in mind, a plethora of work has been done concerning mothers who have continued their exercise regimen throughout their pregnancy. Current research is trying to determine the effects of exercise exposure and its dose on overall fetal growth and development.

L. E. May, *Physiology of Prenatal Exercise and Fetal Development*,
SpringerBriefs in Physiology, DOI: 10.1007/978-1-4614-3408-5_3,
© The Author(s) 2012

The main measures of growth used to understand these effects are birth or body weight, length or height, other morphometric measures, and organ development.

Birth Weight

The most commonly documented neonatal health outcome is birth weight. Results have varied regarding the effects of maternal exercise on birth weight, though differences may be reported the weights are still within normal birth weight range (Sternfeld et al. 1995; Hopkins et al. 2011). To begin with, some research shows no change in birth weight regardless of exposure or not to maternal exercise (Collings et al. 1983; Carpenter et al. 1988; Rose et al. 1991; Hatch et al. 1993; Sternfeld et al. 1995; Horns et al. 1996; Clapp et al. 2000, 2002). Large population-based cohort studies suggest that birth weight is determined mainly by confounding variables, such as maternal body composition, with the influence of maternal exercise as minimal (Fleten et al. 2010; Juhl et al. 2010). Some research reports an increase in birth weight (Collings et al. 1983; Carpenter et al. 1988; Rose et al. 1991; Hatch et al. 1993; Sternfeld et al. 1995; Horns et al. 1996; Clapp et al. 2000, 2002). These studies, however, all had low volume exercise (~ 3 times per week) for the last half of pregnancy (~ 20 weeks to term), though they differed in the amount in the first half: no exercise (Collings et al. 1983); low volume (Clapp et al. 2000); moderate-high volume (Bell and O'Neill 1994; Clapp et al. 2000, 2002). Yes, exercise by the mother has been implied to decrease the risk of a large for gestational age infant. Conversely, some neonates exposed to maternal exercise were normal weight, but lighter than neonates not exposed (Curet et al. 1976; Collings et al. 1983; Clapp and Dickstein 1984; Clapp and Capeless 1990; Bell and Sejnowski 1995; Kardel and Kase 1998; Clapp 2000; Kramer 2000; Marquez-Sterling et al. 2000). These latter studies are associated with a moderate to high volume of exercise (5 days per week or more) in the last half of pregnancy. Importantly, a vast amount of research shows that maternal exercise does not increase the risk of low birth weight (76, (Jarrett and Spellacy 1983; Klebanoff et al. 1990; Rabkin et al. 1990; Rose et al. 1991; Sternfeld et al. 1995; Horns et al. 1996; Schramm et al. 1996; Leiferman and Evenson 2003). Though methodology has differed, there seems to be a dose-response relationship between maternal activity level (i.e., energy expenditure, intensity) and infant birth weight (Hatch et al. 1993) unpublished data). A possible explanation for these findings is that the decrease in birth weight associated with maternal exercise might be a normalization of birth weight in light of the fact that newborn birth weights have increased across generations (Hopkins et al. 2011).

It is likely these differences in findings are due to methodology. For example, each study had different exercise protocols. Secondly, other confounding variables have not been controlled for such as maternal diet, maternal weight gain, socioeconomic status, genetic background (i.e,. father's height), occupational activity, and maternal activity level prior to pregnancy.

As regards the offspring, many studies do not control for gestational age at birth. It is more plausible to support the idea that birth weight determination is multifactorial, including maternal body composition and diet, uteroplacental blood flow and placental transfer of nutrients, and fetal genetics (Phillips et al. 2009). These factors then influence fetal nutrient demand versus placental supply which influence fetal development and the programming of the fetus toward normal or abnormal growth (Phillips et al. 2009).

At the very least, these findings suggest maternal exercise does not adversely affect birth weight. At best, fetal growth as determined by birth weight is within normal range when women exercise at moderate intensity, 3–5 days per week, whereas the extreme of too little or too much exercise, or vigorous intensity in the third trimester leads to changes in birth weight.

To further support the idea that exercise during pregnancy is safe, are the findings that fetal glucose and oxygen delivery is maintained throughout pregnancy even with maternal exercise. Two unique human studies provide more direct evidence of adequate oxygenation in fetuses exposed to maternal exercise (Clapp et al. 1995, 2003; Clapp 2003). Additional studies support the maintenance of appropriate fetal glucose supply (Clapp 2003).

Other Morphometric Measures

Besides birth weight, other morphometric outcomes can be utilized as an indicator of fetal growth during gestation, such as birth length. No differences have been measured in birth length of babies exposed or not exposed to gestational exercise (Clapp 1989; Clapp and Capeless 1990; Botkin and Driscoll 1991; Giroux et al. 2006; Barakat et al. 2009). Although there seems to be a trend of decreasing birth weight and length with increasing maternal exercise, this association is not significant when accounting for confounding variables of gestational age, gender, maternal age, pre-pregnancy BMI, socioeconomic status, parity, and smoking (Juhl et al. 2010). Other measures used to indicate normal fetal growth and development are circumferences and organ development. There are no differences found in measurements of neonatal head circumference, abdominal circumference, and ponderal index, regardless of maternal exercise exposure or not (Clapp and Capeless 1990). Though a large, cohort study found trends toward decreasing offspring size (abdominal and head circumference) in offspring with increasing maternal exercise (Juhl et al. 2010). This interaction was only significant when including smoking as an interaction term (Juhl et al. 2010). Scant research has been done regarding organ development in humans in response to maternal exercise. A study utilizing a rat model finds there are no differences in organ weights (heart, kidney, brain, and liver, skeletal muscle) relative to body weight regardless of maternal exercise exposure (Mottola et al. 1989; Houghton et al. 2000). Similarly, there was no difference in lung volume to body weight in offspring regardless of in utero exercise (Nagai et al. 1993). Rat pups of exercised mothers had increased density of cardiac muscle fibers and capillaries, with decreased diffusion distance relative to pups of non-exercised mothers (Parizkova 1975, 1979). As regards humans, there are decreased risks of congenital abnormalities with maternal exercise exposure (Clapp 1989, 1991; Juhl et al. 2010). Especially for women who are obese, maternal activity helps prevent negative outcomes for the infant (Dye et al. 1997). The risk of neural tube defects, for instance, is decreased about 50% due to maternal exercise during gestation (Carmichael et al. 2002). Overall, data show that maternal exercise does not negatively impact growth and development, nor organ development.

Postnatal Growth

Some researchers suggest that although measurements are proportional, there is still the potential for adverse outcomes later. In the case of lighter babies due to maternal exercise, will there be risks in a "catch up" phase of growth in early childhood. The converse argument is their risk with "catch down" phase in those that were heavier at birth. The data to date show that this is not the case, these infants do not seem to have a "catch up" or "catch down" phase. The offspring of women who exercise demonstrated normal growth and development during the first year of life. (Clapp 1991; Clapp et al. 1998). At 1-year old, there were no differences between children exposed or not exposed to exercise in utero in growth measures: height, weight, head circumeference, chest circumference, abdominal circumference (Clapp et al. 1998). Similar findings were seen in children 5 years of age (i.e., height, and circumferences: arm, head, chest, abdominal) (Clapp 1996). Futhermore, children exposed to exercise in utero have normal blood pressure (systolic, diastolic, and mean arterial) and body morphometry (fat mass, head and abdominal circumference) at 5 years of age (Clapp and Lopez 2007).

Future Research

Future research should account for other variables which may influence fetal growth and development. For example, maternal nutrition status throughout pregnancy may impact fetal growth measures. Furthermore, maternal activity level prior to pregnancy or her fitness level may influence these results of fetal as well as postnatal measures. More studies should also analyze the affect, if any, of maternal exercise on fetal organ development. With current technology, such as ultrasound, MRI, CT, and PET scans, this information is attainable and would be helpful to further understand the influence of maternal exercise on offspring development. Further analysis should be done to determine whether in utero exposure to exercise is associated with the prevention of hypertension and obesity in children. Considering the obesity pandemic and the high mortality related to cardiovascular related diseases, pregnancy research should be a prime focus for prevention research.

Chapter 4
Effect on Postnatal Health and Beyond

Abstract Current evidence suggests regular maternal physical activity while pregnant does not negatively influence in utero or postnatal growth and development. Weight and length/height, however, are fairly crude measures of overall health and well-being. Similar to an adult exercise trained response, infants of women who exercised have improved cardiac ANS relative to same age counterparts not exposed to maternal exercise. Furthermore, adults who were exposed to exercise in utero had no signs of cardiovascular disease. Offspring exposed to maternal exercise show evidence of improved nervous system development as suggested by academic and sports performance even into adulthood. This is also seen on aspects of intelligence tests, such as language scores. Children of women who exercised during pregnancy showed evidence of better temperament. Further, children exposed to exercise in utero had decreased adiposity at birth and into adulthood. These postnatal findings have enormous implications on conditions such as cardiovascular disease, hypertension, obesity, diabetes. Considering the potentially positive impact on public health, more research should be done in this area.

Keywords Prenatal programming · Cardiovascular system · Nervous system · Intelligence · Neurobehavioral · Adiposity

There is sufficient evidence to conclude that adult health is partly programmed in utero. This statement does not imply that decisions and environment in adult life is not a factor in an individual's health, though it may explain why adult risk factors predict disease so poorly. (Barker 1997a, b). With the explosion of research related to fetal or prenatal programming, many disciplines have studied the association between adult disease and in utero environment. Though this research has grown exponentially, many studies still rely on birth weight, or similar data. Although birth weight data is the most available and easiest to acquire, it does not provide a complete picture about how various changes influence the gestational environment (DiPietro et al. 2010). This chapter will move beyond the initial neonatal measures.

L. E. May, *Physiology of Prenatal Exercise and Fetal Development*,
SpringerBriefs in Physiology, DOI: 10.1007/978-1-4614-3408-5_4,
© The Author(s) 2012

The developmental origins hypothesis states that the in utero environment influences fetal development and further contends that these differences will have implications for the child throughout their life span into adulthood. Overwhelming research suggests that continuing or beginning exercise while pregnant does not have adverse effects on the offspring. Furthermore, continuing evidence suggests that participating in activity during pregnancy may have beneficial effects for offspring.

Cardiovascular System

Previous data (presented in Chap. 2) demonstrated exposure to regular maternal exercise alters the cardiac ANS in ways similar to an adult exercise trained response. Furthermore, persistence of heart rate and HRV within individuals is well established (Bornstein et al. 2002; DiPietro et al. 2002a, 2004, 2007), as well as how these measures relate to developmental outcomes.

Similar to findings in the fetus (Van Leeuwen et al. 2003; David et al. 2007; May et al. 2010) 1-month-old infants exposed to regular Leisure time physical activity (LTPA) in utero had lower HR and increased HRV compared to infants not exposed to maternal LTPA (May et al. 2011a, b). HRV was found to increase up to 1 month of age (Finley and Nugent 1995; Nagaoka et al. 2003). The persistence of lower HR and increased HRV from the fetus until after birth agrees with DePietro findings (DiPietro et al. 2007) and supports the premise of prenatal programming (Barker 1997a, b). Additionally, investigators found a dose of maternal LTPA relationship with the infant cardiac measures, similar to what is observed in utero (May et al. 2008). In addition, offspring exposed to maternal exercise in utero exhibited no evidence of cardiovascular disease up to 17–20 year of age (2006). Therefore, a regular physical activity program during pregnancy provides cardiovascular health benefits for the baby after birth. This positive lifestyle can be an early intervention to attenuate, or alleviate chronic conditions, such as cardiovascular disease and diabetes. Further, these findings support the idea of fetal programming during fetal life and the association with birth weight and coronary heart disease (Barker 1993, 1997a, b). Fetal programming can be associated with positive postnatal outcomes. This cardiac autonomic programming of offspring may potentially increase quality of life and reduce healthcare costs.

Nervous System

Since the heart is controlled by the ANS, lower HR and increased HRV indicates a more mature nervous system development as well. For example, slower HR and increased HRV in the fetus are associated with significantly higher Motor Development Indices, and better language development at 2 years of age (DiPietro et al. 2007). Outcomes for children aged 8–12 years indicate similar growth, significantly better academic and extracurricular performance, and equal or advanced coordination, balance, strength, speed, and endurance over children not exposed to maternal exercise. Clapp reported that 17–20 year olds who were exposed to maternal exercise had significantly better academic and sports performance. These findings agree with research by Porges (1972; Porges and Humphrey 1977) showing study participants with increased HRV had faster reaction times and attention during a task.

Intelligence

The utility of prenatal cardiac indicators in predicting postnatal development was studied in a group of low-risk, non-smoking women. (DiPietro et al. 2007) FHR and HRV monitoring was performed during gestation then child development was evaluated at 2 and 2.5 years of age using the Mental Development Index (MDI) and Psychomotor Development Index (PDI) of the Bayley Scales of Infant Development II (BSID II). The study found that fetal HRV was significantly and positively associated with child development during the third year of life. Fetuses exhibiting slower and more variable heart rates had significantly higher MDI scores, and significantly better language development than those with faster and more stable heart rates, suggesting that there may be conservation of developmental rate from the prenatal to the postnatal period. The findings of this study were consistent with another study that explored the relationship between neonatal physiological measures, such as mean heart period and cardiac vagal tone, and school-age (6–9 years) outcome measures. Significant associations were found between vagal control and mental processing scores, and gross motor development 3 years later (Doussard-Roosevelt et al. 2001). These studies support the hypothesis that prenatal regulation of the autonomic nervous system is predictive of complex behaviors later in life.

Interestingly, Clapp et al. found that 5-year-old children exposed to maternal exercise in utero had higher intelligence scores (Wechsler test scales) and language skills than 5-year old counterparts who were not exposed to maternal exercise in the womb (Clapp 1996). This agrees with previous research demonstrating positive developmental outcomes (i.e., Bayley Scales of Mental Development) associated with increased vagal tone (slower HR and increased HRV) in late gestation (Fox and Porges 1985). Further, a study found an association between increased fetal HRV, similar to that in exercise-exposed fetuses, and neonatal mental, psychomotor, and language development at 3 years of age (DiPietro et al. 2007). These findings further support the idea that improved measures of physiological processing, seen as increased HRV due to gestational exercise exposure, are an index of central nervous system integrity and augmented cognitive processing (Fox and Porges 1985).

Neurobehavioral/Temperament

The offspring of 34 women who exercised during pregnancy were compared with those of 31 control subjects. Behavior of the infants was assessed 5 days after birth using the Brazelton Neonatal Behavioral Assessment Scales (NBAS). The offspring of the exercising women scored higher in 2 of the 6 behavioral categories examined. The scores for habituation, motor organization, autonomic stability, and behavioral state range were not significantly different between the groups, but the scores reflecting ability to regulate state (i.e., quiet self) after stimuli and ability to orient to environmental stimuli were significantly higher in the exercising group. Researchers concluded that the neurological development of neonates who are born to exercising mothers may be more mature as early as the fifth day after birth, than neonates of born to sedentary mothers. These data supported the hypothesis that continuing regular exercise throughout pregnancy alters early neonatal behavior in a positive direction. The changes in ability to regulate state and ability to orient to environmental stimuli, in particular, were consistent with learned responses to various stimuli associated with exercising, such as rhythmic motion, which would be detected by the fetus in utero (Clapp et al. 1999).

Adiposity

Although many morphometrics were similar in neonates from mothers who exercised during pregnancy compared to those who did not exercise, children were lighter at 5 years of age relative to children from women who did not exercise while pregnant (Clapp 1996). This difference in body weight was explained by less fat mass (Clapp 1996). Further research has looked at body fat in prepubertal children and the factors associated with the intrauterine programming of such. The percent of body fat is inversely related to birth weight, after adjusting for child's BMI, age, sex, current physical activity, and socioeconomic status (Elia et al. 2007). These findings suggest that 4% more body fat in prepubertal children occurs in the 10th percentile of birth weight relative to the 90th percentile of birth weight (Elia et al. 2007). In addition, offspring exposed to maternal exercise in utero exhibited no evidence of obesity into their twenties (2006). Considering the current obesity epidemic these results have serious implications for public health.

Breastfeeding

Little research has been done related to the effect of exercise, breastfeeding, and neonatal development in the postpartum period. To date, it is known that exercise does not affect maternal supply or general composition of milk (Lovelady et al. 1990; Dewey and McCrory 1994; Prentice 1994). However, lactic acid is present in milk after an exercise bout (Wallace et al. 1992a, b; Carey et al. 1997, 1998;

Quinn and Carey 1999). Babies of exercising mothers who are breastfed experience normal growth, even if the mother continues to lose weight in the postpartum period (Dewey and McCrory 1994). Based on current findings, exercising in the postnatal period is not detrimental to the mother's ability to produce milk, nor in the nutrient supply for the child.

Conclusion

Altogether these findings in offspring during and after pregnancy support improved health, and indicate Peripheral and Central Nervous System development. The research concerning how exercise interacts with fetal development on a long-term basis still leaves much to be discovered. Studies have shown that there is conservation of developmental rate from the prenatal to the postnatal period, and that measurements such as the combination of FHR and HRV serve as indicators of individuality in developing neural control during pregnancy and have predictive value for postnatal functioning (16). Data from a relatively small study compared a group of exercising mothers to a group of mothers who refrained from all physical activity besides walking during their pregnancy. This study found that the offspring of the exercising group scored significantly higher on tests of general intelligence and oral language skills at 5 years of age. The authors suggest that it is possible that some stimuli, such as vibration, sound, motion, etc., associated with exercising may affect the neurodevelopment of the fetus in a beneficial way. (Clapp 1996) It seems to be well established that the offspring of women who perform certain types of exercise during pregnancy have improved morphometric and neurobehavioral differences which are beneficial (21), in addition to the numerous advantageous effects for the mother.

Future Research

Considering the exciting findings thus far linking maternal physical activity and positive long-term outcomes, further carefully controlled research is warranted in this area. In order to verify these initially encouraging findings, research needs to be done in regard to postnatal outcomes. For example, considerations of maternal activity level prior to pregnancy should be measured and controlled. Additionally maternal nutrition during pregnancy as well as during breastfeeding must be measured as well. Additionally, maternal pre-pregnancy BMI and weight gain, and fat gain during the pregnancy should be accounted. Other factors of maternal consideration are genetic background, smoking, diabetes, and socioeconomic status. Additionally, the environment enrichment after birth (i.e., education, healthy eating habits, physical activity encouragement) should be controlled for as well in order to more clearly define the effect of maternal exercise during pregnancy on long-term benefits to the child.

Chapter 5
What are the Barriers to and What is OK for an Exercise Study

Abstract Despite growing evidence supporting the safety of physical activity during pregnancy and possible benefits, there are many women who do not stay active or become active during pregnancy. There are many reasons women do not participating in activity during gestation. Understanding these reasons is important for health-care providers who should be encouraging women to stay active while pregnant. Furthermore, it is important for providers to understand what motivates pregnant women to change their behavior. In order to effectively intervene regarding activity and pregnancy, providers must know the current guidelines regarding safe practices for exercising while pregnant. Current recommendations for aerobic and strength training are provided. Considering the potential gains for mother and child, it is imperative for pregnant women to know and understand what activities can be done during pregnancy and how it affects pregnancy outcomes.

Keywords Barriers · Health-care provider · Aerobic · Strength training

A large number of pregnant women are at increased risk of pregnancy complications due to excess weight, diabetes, and hypertension. Although these conditions can be attenuated or alleviated by exercise, in the US, less than a quarter of pregnant women meet the recommended guidelines for activity during pregnancy (Petersen et al. 2005; Evenson and Wen 2010a, b). A similar lack of maternal activity is found outside of the US as well (Higgins et al. 2011). Overall, women participated in more activity during the first trimester compared to the third trimester (Schramm et al. 1996; Evenson and Wen 2010a, b). Since there are benefits of being active throughout the entire gestation, there is great value in determining the barriers and motivations to exercise during pregnancy (Clapp and Dickstein 1984; Malnick and Knobler 2006; Siega-Riz and Laraia 2006; D'Angelo 2007).

L. E. May, *Physiology of Prenatal Exercise and Fetal Development,*
SpringerBriefs in Physiology, DOI: 10.1007/978-1-4614-3408-5_5,
© The Author(s) 2012

Barriers to Exercise

Identifying reasons pregnant women use to justify lack of physical activity may help improve the efficacy of exercise intervention strategies and ultimately improve the health of mother and fetus. Numerous studies have noted various barriers to women not being involved in physical activity while pregnant. The main reason given for choosing not to exercise during pregnancy was lack of time (Smith et al. 2008). Other reasons for not exercising include tiredness, physical discomfort, other children (i.e., child care and time), and not knowing what to do (Smith et al. 2008; Evenson et al. 2009; Mudd et al. 2009; Evenson and Bradley 2010). Among overweight and obese pregnant women, barriers also include physical and psychological barriers, work, and environmental factors (Weir et al. 2010) in addition to family commitments and time.

Despite reports that lack of time is the most common reason for not exercising during pregnancy, the daily activity data suggest women have time to exercise. Recent research revealed that outside of their exercise routine, exercisers spend either the same amount, or more time than nonexercisers performing daily activities (May and Gustafson 2009). It appears that women who fit exercise into their schedule are more effective at managing their time. Physicians should continue to encourage pregnant patients to exercise during pregnancy and possibly consider introducing time management skills as well. Additionally, health-care providers should assist their overweight and obese pregnant patients find programs to assist them beginning and maintaining an exercise routine throughout pregnancy and after birth.

Similar to previous research, one study found that pregnant women lead active lifestyles but may not engage in regular exercise (Chasan-Taber et al. 2004). Most of the reasons given for choosing not to exercise during pregnancy can be addressed through provider–patient interaction and education. For women who do not know what to do, who were afraid, who may not like exercise, were not sure why they did not exercise, or were concerned about transportation or money for a gym. Providers can educate and encourage women on the breadth of exercise activities that are safe and accessible during pregnancy (Clapp et al. 1999, 2002). Other investigations show that for overweight and obese pregnant women, behaviors will not change without education and intervention from a health-care provider (Mottola 2009). For women who reported feeling too tired, sick, or in pain to exercise, providers can explain how exercise during pregnancy is known to provide more energy, decrease musculoskeletal pain, enhance mood and sense of well-being, and possibly result in shorter labor and delivery (Clapp 1994) , and improve sleep (May and Gustafson 2009). Exercise intervention may be especially helpful to overweight and obese pregnant women since they often have more symptoms and complications associated with pregnancy (Hall and Kaufmann 1987). For most of these reasons, education and instruction on time efficient ways to increase activity may help increase participation since exercise is known to decrease or alleviate these barriers.

Other studies have searched for reasons why women would decide to exercise during pregnancy. First and foremost, many women are more motivated to be healthier in general during pregnancy. In overweight and obese women, they are further motivated by a desire to minimize weight gain (Weir et al. 2010). However, many women expressed beliefs about exercise and possible negative pregnancy outcomes (Weir et al. 2010). For example, pregnant women polled report that moderate activity while pregnant is safe, but they are not informed concerning vigorous activity participation (Duncombe et al. 2009; Mudd et al. 2009).

Women who chose to exercise while pregnant spent more time performing various sedentary and non-sedentary daily activities overall than nonexercisers (Smith et al. 2008). Although one study found no demographic difference in women who did or did not exercise while pregnant (Smith et al. 2008), other studies have found women who are non-Hispanic White ethnicity, nulliparous, have greater than a high school education, are older than 25 years of age, do not smoke, and were active before conception, are more likely to exercise during pregnancy (Petersen, Leet et al. 2005; Fell, Joseph et al. 2009). Understanding these demographics are useful to appropriately target interventions.

Throughout pregnancy, women receive vast amounts of information some of which is conflicting regarding activity during their pregnancy (Clarke and Gross 2004). Thus, the individuals who provide various counsel to women may be influential in their decisions to exercise during pregnancy. Health-care provider discussion of exercise is related to women choosing to exercise throughout gestation (unpublished data; (Schramm, Stockbauer et al. 1996). These data agree with other studies showing that health-care provider intervention can change patient behavior which can improve pregnancy outcomes (Brooten, Youngblut et al. 2001; Clothier, Stringer et al. 2007; Radnai, Pal et al. 2009). Furthermore, patients' compliance associated with behavioral changes is highest when there are frequent contacts (Joshi 1999; Brooten, Youngblut et al. 2001). Though health-care provider intervention throughout gestation may be more effective than other times due to the frequent nature of prenatal visits, this does not explain why so little women exercise during pregnancy. This may be explained by the fact that less than three-quarters of physicians discuss prenatal exercise (Schramm, Stockbauer et al. 1996; Krans, Gearhart et al. 2005). Also, the majority of OB/Gyn physicians are not aware of the current ACOG guidelines (Bauer et al. 2010). Due to this situation, many physicians advise women to decrease activity or do less than the current recommended guidelines (Schramm, Stockbauer et al. 1996; Krans, Gearhart et al. 2005). Additionally, providers must be sure to address women's concerns related to the safety and efficacy of activity during pregnancy. Therefore, health-care providers should be familiar with current guidelines, and also be ready to instruct on what is safe and how to exercise while pregnant.

Women are often likely to believe the outdated claims or myths from magazines, family, and friends instead of their physicians or exercise professional (Clarke and Gross 2004). As heart disease and obesity-related conditions continue to elicit concern in our society, it is crucial that health-care providers educate pregnant women on safe exercises during pregnancy (Smith et al. 2008).

Current Guidelines Regarding Physical Activity During Pregnancy

Another important step in targeted interventions focuses on knowing current guidelines. The following paragraphs will describe guidelines according to the American Congress of Obstetrics and Gynecology (ACOG), American College of Sports Medicine (ACSM), Canadican Society for Exercise Physiology (CSEP), Centers for Disease Control and Prevention (CDC), and Society of Obstetricians and Gynaecologiests of Canada (SOGC). These guidelines will cover who should exercise, when to start, type of exercise, frequency, intensity, and length of exercise regime.

Who can Exercise While Pregnant

Regardless of previous activity level, all women free from pregnancy complications should participate in aerobic and strengthening exercises (CSEP, SOGC, ACOG). These professional organizations have similar absolute and relative contraindications to exercise during pregnancy. Specifics from the Canadian organziations CSEP and SOGC are outlined here, as well as current guidelines from ACOG and ACSM (2002; Davies et al. 2003). Prior to participation, women should complete a questionnaire, such as the Physical Activity Readiness Medical Examination for Pregnancy (PARmed-X for Pregnancy). This free readiness questionnaire can be found at (found at http://www.csep.ca/forms.asp or in GET-P8 (Kaufman et al. 2006). After completing a readiness form and speaking with a health-care provider, women can begin, or continue, an exercise routine. However, women as well as those who advise pregnant women should familiarize themselves with the signs to terminate exercise: bleeding, dyspnea, dizziness, headache, chest or calf pain/swelling, muscle weakness, preterm labor, decreased fetal movement, or amniotic fluid leakage ACOG Committee #267. Other reasons for termination of an exercise session are excessive shortness of breath, presyncope, painful uterine contractions (Davies et al. 2003).

When to Start Exercise During Pregnancy

Many women and health-care providers do not know the most appropriate time to initiate an exercise protocol. Women who were previously active can continue their exercise routine, aerobic and strength exercises, as soon as they learn of their pregnancy, as long as they are symptom free. Those women who were previously sedentary can begin light activity anytime during pregnancy. However, some women prefer to wait until after the first trimester is completed due to feelings of nauseousness and tiredness (Davies et al. 2003).

Aerboic Exercise for Pregnant Women

Frequency and Duration

Important aspects of exercise during gestation to discuss are factors involving frequency, how often during the week, and duration, how long each session lasts, intensity, of maternal exercise. Most physicians believe that 3–5 exercise sessions per week are ideal (Stevenson et al. 1997). For women who have not previously exercised, ACOG recommends starting slowly, and building up to 3 sessions per week (2002). For women who have been active prior to conception, guidelines recommend women exercise at least three per week, or everyday of the week (2002). The amount of time per session may vary from 20 to 60 min, or more, depending on the activity. Some guidelines report the suggested amount of time per week, such as total minutes per week. For women who were not previously active, guidelines recommend 90 min per week (about 30 min, 3 times per week). For women who were sedentary prior to conceiving, it is suggested they begin with 15 min of comfortable activity 3 days of the week. Every week she can add 5 min to her daily total as long as she can talk comfortably while exercising and has no pain or symptoms. Once she reaches 30 min per session, an additional day can be added or time each day, if desired. Otherwise, guidelines for healthy already active women are 150 min per week (i.e., 50 min for 3 days, or 30 min for 5 days (Mottola 2009)). Pregnant women should always precede and succeed the aerobic session with a brief warm-up/cool-down session (i.e., light stretching, slow walk).

A large retrospective cohort study found first trimester aerobic exercise of ≥75 min per week could be associated with an increased risk of miscarriage (Madsen et al. 2007). During the first trimester of pregnancy, the ideal exercise time ranged from 45 to 74 min per week; as a retrospective study though recall bias may have influenced these values. In the second and third trimesters there were no associations with miscarriages and amount of exercise (Madsen et al. 2007). Additionally, one study suggests too much exercise (>5 times per week), similar to not enough (≤2 times per week), may be associated with an increased likelihood of delivering a small for gestational age baby (Campbell and Mottola 2001). Though these findings have not been verified by other studies, small for gestational age is a risk factor for obesity and cardiovascular disease later in life (Campbell and Mottola 2001). It seems that the safest frequency of an exercise program should be 3–5 times per week (Campbell and Mottola 2001). Most pregnant women can exercise most days of the week, for about 30 min per session.

Intensity of Aerobic Activity

ACOG and ACSM guidelines recommend pregnant women to participate in this amount of activity at moderate to vigorous intensity, in the absence of pregnancy complications. There are three main ways to estimate intensity while exercising: Borg's rating of perceived exertion, target heart zones, or the "talk test." The first intensity determination is the Borg's rating of perceived exertion and is a scale from 6 (extremely light) to 20 (extremely hard). During gestation, a rating of 12–14 RPE is considered appropriate (Davies et al. 2003). Although competitive and highly trained athletes might safely train harder during pregnancy than most women, for the average fit individuals an RPE of 12–14 (moderate) is suggested, but this may decrease in late pregnancy. Secondly, a more specific modified target heart zone can be used for aerobic exercise during pregnancy. This chart is based on work by Mottola and defines target heart zones for each maternal age category (Kochan-Vintinner et al. 1999 booklet http://www.scep.ca/publicationsmain.html): <20 years HR target is 140–155 bpm; 20–29 years HR target is 135–150; 30–39 years HR target is 130–145; and >40 years HR target is 125–140. Lastly, the talk test is a term described to estimate intensity based on the women's ability to converse during her exercise routine. If the exercise intensity is comfortable, then she should be able to talk during exercise. If she is not able to comfortably talk, then her intensity should be reduced. In order to determine a comfortable intensity, women can use the visual Borg scale (12–14 RPE), a physical HR measurement (target HR), or the general talk test to estimate her intensity level during each exercise session.

Type of Aerobic

Once appropriate frequency, intensity, and time are established, the mode(s) of aerobic exercise is chosen. Aerobic exercises which are non-impact and non-weight bearing, such as swimming and stationary cycling are the safest during all trimesters of pregnancy. Research has demonstrated that swimming is the safest aerobic activity throughout pregnancy, due to the low impact nature and improved thermoregulation (Madsen et al. 2007). However, swimming and stationary cycling require a gym membership or costly equipment which is not feasible for many pregnant women. Walking is a low-impact exercise which is the most frequently used during pregnancy; most likely since no special equipment is required, just good shoes, it can be done anywhere and at any intensity, and it is well tolerated throughout the pregnancy. Although jogging and running are moderate and high intensity exercises, they have also been shown to be safe during pregnancy, in low risk pregnancies. Throughout the spectrum of exercise intensities and gestation, adverse outcomes have not been associated with maternal activity level and pregnancy outcomes.

During pregnancy some exercises must be avoided while others only need modifications. As the pregnancy progresses, the uterus and the growing fetus expand out from the protective pelvic cradle into the open abdominal area. Though the risk of abdominal injury is very low (Finch 2002) and there are have been no adverse pregnancy outcomes (Clapp 2000) related to exercise, it is still advisable to modify or avoid some activities. Sports that involve fast moving balls or objects should be avoided (i.e., ice hockey, ball sports, court sports, etc.). Women should also exclude those activities which involve speed or potential for injury or falling (i.e., gymnastics, horseback riding, water skiing, martial arts, etc.). Sports that can be mofidied include outdoor skiing and bicycling. ACOG recommends snow skiing on safe slopes only. Outdoor bicycling should be modified to stationary cycling or spinning, while outdoor skiing can be switched to indoor cross-country skiing.

Women who were active prior to conception may maintain their regular aerobic regimen, as long as their current exercise is not on the activities to be avoided list. Beginners to exercise and pregnancy should choose a low to moderate impact, fun activity. Low and moderate impact aerobic activities include swimming, walking, jogging, spinning/cycling, aerobics classes, and most aerobic equipment (i.e., stair climber, elliptical, rowing). Avoid activities that may decrease maternal circulation by compressing the inferior vena cava (i.e., Crossrobics). Late pregnancy hormonal changes causing joint laxity may necessitate further modification of aerobic activity, such as in rowing. Studies have also documented the effects of aerobic exercise as well as anaerobic strength and conditioning activities.

Strength Training During Pregnancy

Though the ACOG and ACSM do not mention strength exercises specifically, the SOGC and the CSEP recommend women in uncomplicated pregnancies participate in aerobic and strength-conditioning exercises (Davies et al. 2003). The lack of detail as it relates to resistance exercises during pregnancy is a result of the amount of research in the area. Most of the focus has been on aerobic exercise and much less has been done to date relating to the effects of strength exercises while pregnant.

Duration and Types of Strength Training

Since there is less research regarding this area, a conglomeration of the general routine and types of training will be presented. As with any exercise program, there should always be time for proper warm up to prepare the muscles for activity, the training program, followed by a cool down and stretching time. To warm up for strength or any exercise, women should take 5–10 min of slow walking or light cycling (Hall and Kaufmann 1987; Barakat et al. 2009a; Oostdam et al. 2009; de Barros et al. 2010; O'Connor et al. 2011). The middle and main component of the session is the strengthening session. This portion should last from 20 to 45 min, variations will be dependent on the number of exercises performed which can vary from 4 to 12 (Hall and Kaufmann 1987; Barakat et al. 2009a; Oostdam et al. 2009; de Barros et al. 2010; O'Connor et al. 2011). Women can do 1–3 sets of each exercise, and all studies suggest repetitions of 8–15 (Hall and Kaufmann 1987; Barakat et al. 2009a; Oostdam et al. 2009; de Barros et al. 2010; O'Connor et al. 2011). Women should use slow and controlled movements, and make sure to breathe throughout the full range of motion (Barakat et al. 2009a; O'Connor et al. 2011). At the end of the Strengthening program should be the cool down and stretching period. This session should last about 5–10 min (Barakat et al. 2009a; Oostdam et al. 2009) to allow for maternal heart rate to return to pre-exercise levels. The best activity for this cooling period is slow walking or slow, steady stretching of major muscle groups.

All strengthening protocols should train the major muscle groups: quadriceps, hamstrings, back, chest, deltoid, triceps, biceps, and calf muscles (Hall and Kaufmann 1987; Barakat et al. 2009a; Oostdam et al. 2009; de Barros et al. 2010; O'Connor et al. 2011). Some studies included core training with resistance training. Although this area is overlooked during training, it is essential for the pregnant exerciser. The core muscles (i.e., abdominals and back muscles) are important in maintaining posture, especially with her changing center of gravity. Furthermore, the core is also essential during the labor and delivery process. Strengthening and abdominal exercises can be done at an angle (inclined), on the side, seated, or standing. There are many forms of resistance that have been used during pregnancy as well. The safest is using a woman's own body weight, as in push-ups or planks, to do resistance training. Other ways to resistance training are by using resistance bands or similar products, weight machines, circuit training, or even free weights and barbells. Heavier weights and resistance may be fine in women who are athletes and/or already exercise with heavy weights prior to pregnancy (Hall and Kaufmann 1987; Barakat et al. 2009a; Oostdam et al. 2009). At all times it is imperative that women use slow and controlled movements and avoid any movements that may cause blunt abdominal trauma.

Intensity and Frequency

Although resistance exercise is not always a continuous intensity like aerobic exercises, similar rules can apply. Though the talk test, and target HR can be used, due to the intermittent nature of strength exercises, often the Borg RPE scale is used. Intensity of exercise should be checked throughout the session. On the 6–20 RPE scale, pregnant women should aim for a moderate intensity level, about 13 or somewhat hard (Hall and Kaufmann 1987; Barakat et al. 2009a; Oostdam et al. 2009; de Barros et al. 2010; O'Connor et al. 2011). It is suggested that women perform resistance exercise programs about 2–3 times per week (Hall and Kaufmann 1987; Barakat et al. 2009a; Oostdam et al. 2009; de Barros et al. 2010; O'Connor et al. 2011).

General Considerations During Pregnancy and Exercise

Considering the change in a woman's body during pregnancy and to maintain a safe physiological environment for mother and fetus, a few other items are worth mentioning. Exercise normally involves an increase in body temperature. Although the maternal physiology has numerous adaptations to maintain a heat gradient away from the fetus, it is still important for women to train in an environment that preserves this gradient away from the fetus. Therefore, women should be and stay well hydrated and activity should be done in a comfortable environment, not too hot, too humid, nor too cold. In order to maintain appropriate blood flow, oxygen, and nutrients to working muscles and the fetus, women should not wear clothing that is restrictive. Additionally, they should not lay supine or prone after 13 weeks gestation. It is best for the woman to sit upright or at an angle, or stand. It is also advised that she eat a healthy, light snack 1–2 h prior to her activity. Certain aerobic activities are intrinsically jolting in nature. Though this does not cause a problem with the placenta or fetus, women should wear a supportive bra (not a sports bra) to protect the enlarging breast tissue. The shift in the center of gravity due to an enlarging abdominal area may require support for comfort. Therefore, a belly band or sling can be worn during their exercise activity. Ideally, women should feel comfortable and pain-free while exercising.

Increasing aerobic and strength activities is safe during pregnancy. Research is still being done in this area to understand women's perceptions and participation in exercise while pregnant. It is imperative that health-care providers are educated on current guidelines to help educate women on what is safe and beneficial during their pregnancy. The current recommendations for frequency, intensity, duration, and type of aerobic as well as strength training participation should be provided to all pregnant women. Just as important is the accurate, and up-to-date, information about activities that should be avoided during pregnancy as well as those activities that need to be modified during pregnancy. Additionally, women and their health-care providers should discuss who should not participate in exercise while pregnant, and the signs for immediate termination of an exercise session. Overall, much research is being done regarding exercise during pregnancy, but more is needed.

Chapter 6
Conclusions

To summarize, there is no evidence to suggest that exercise during pregnancy causes fetal distress during or after maternal exercise bouts. Collectively, there is evidence during pregnancy of beneficial alterations in fetal development resulting from maternal exercise. These differences can be seen in fetal heart rate, breathing movements, and motor activity. Together, these findings reflect maturation of the fetal nervous system. As it relates to the date of delivery, maternal exercise influences this critical time as well. Maternal exercise is shown to either have no effect or improve outcomes of laboring, delivery, and neonatal well-being. After the baby is born, measurements can be done to determine the influence of gestational exercise on overall fetal growth and development. Studies show that there are no adverse effects on overall fetal growth as evidenced by length, weight, and circumference measures. There is some evidence to suggest that exposure to maternal exercise may be beneficial as it relates to decreased fat, possibly decreasing the susceptibility to obesity later in life. Beyond birth, findings are similar. As regards cardiovascular development, exposure to maternal exercise seems to lead to more mature cardiac autonomic nervous system infants and children. This may be cardioprotective from future cardiac disease and hypertension. For growth measures there are no differences, except for the potential for decreased fat mass into childhood. Again, the inclination is that exposure to gestational exercise may help attenuate the risk of obesity. There is also support for improved maturation of the child's nervous system.

Altogether these findings are informative and suggest that activity during pregnancy may be the earliest intervention to improve offspring cardiovascular health, body morphometry, and nervous system development. Ultimately, these data suggest a potentially positive prenatal programming effect, which implies that pregnancy should be the first focus for interventional strategies to help promote health and reduce the risk of cardiovascular disease, hypertension, and obesity later in life. These findings indicate that the in utero environment is enhanced from maternal exercise. As a result, fetal development is influenced in a manner in which offspring are healthier which may result in decreased health-care costs.

L. E. May, *Physiology of Prenatal Exercise and Fetal Development*,
SpringerBriefs in Physiology, DOI: 10.1007/978-1-4614-3408-5_6,

Therefore, health-care providers should encourage their pregnant patients to participate in regular exercise throughout their pregnancy. Moreover, they support all women with uncomplicated pregnancies to engage in activity while pregnant based on current ACOG, ACSM, CSEP, and SOGC recommendations (2002). These guidelines address when to start, type of exercise, frequency, intensity, duration of exercise regime, as well as modifications necessary, and environmental considerations.

Further research, especially longitudinal studies, is necessary to further substantiate these initial findings and long-term implications. Additionally, more tightly controlled studies are needed to determine possible confounding variables as well as associations. Also, ways to standardize exercise volume, intensity, and workload during pregnancy will help with future research findings as well as to better interpret previous findings. Eventually, these data should be linked to savings in health-care costs.

Ultimately, the initial findings have substantial implications on public health in many aspects of child and long-term health. Considering the tremendous impact this focus of research has on maternal, child, and adult health, public health across generations, and health-care costs, more research funds should be focused on this exciting area of research with long term public health implications.

References

ACOG Committee Opinion (2002) Number 267, January 2002: exercise during pregnancy and the postpartum period. Obstet Gynecol 99(1):171–173

American College of Sports Medicine (2010) ACSM's guidelines for exercise testing and prescription. Lippincott Williams & Wilkins, Philadelphia

Artal R, Fortunato V et al (1995) A comparison of cardiopulmonary adaptations to exercise in pregnancy at sea level and altitude. Am J Obstet Gynecol 172(4 Pt 1):1170–1178 (discussion 1178–1180)

Avery ND, Stocking KD et al (1999) Fetal responses to maternal strength conditioning exercises in late gestation. Can J Appl Physiol 24(4):362–376

Ayres-de-Campos D, Bernardes J (2004) Comparison of fetal heart rate baseline estimation by SisPorto 2.01 and a consensus of clinicians. Eur J Obstet Gynecol Reprod Biol 117(2): 174–178

Ayres-de-Campos D, Bernardes J et al (2004) Can the reproducibility of fetal heart rate baseline estimation be improved? Eur J Obstet Gynecol Reprod Biol 112(1):49–54

Bailon R, Mateo J et al (2003) Coronary artery disease diagnosis based on exercise electrocardiogram indexes from repolarisation, depolarisation and heart rate variability. Med Biol Eng Comput 41(5):561–571

Barker DJ, Osmond C, Simmonds SJ, Wield GA (1993) The relation of small head circumference and thinness at birth to death from cardiovascular disease in adult life. Bmj 306(6875): 422–426

Barakat R, Lucia A et al (2009a) Resistance exercise training during pregnancy and newborn's birth size: a randomised controlled trial. Int J Obes 33(9):1048–1057

Barakat R, Ruiz JR et al (2009) Type of delivery is not affected by light resistance and toning exercise training during pregnancy: a randomized controlled trial. Am J Obstet Gynecol 201(6):590 e591–590 e596

Barker DJ (1997a) Intra-uterine programming of the adult cardiovascular system. Curr Opin Nephrol Hypertens 6(1):106–110

Barker DJ (1997) Intrauterine programming of coronary heart disease and stroke. Acta paediatrica 423:178–182 (discussion 183)

Bauer PW, Broman CL et al (2010) Exercise and pregnancy knowledge among healthcare providers. J Womens Health 19(2):335–341

Bell AJ, Sejnowski TJ (1995) An information-maximization approach to blind separation and blind deconvolution. Neural Comput 7(6):1129–1159

Bell R, O'Neill M (1994) Exercise and pregnancy: a review. Birth 21(2):85–95

L. E. May, *Physiology of Prenatal Exercise and Fetal Development*,
SpringerBriefs in Physiology, DOI: 10.1007/978-1-4614-3408-5,
© The Author(s) 2012

Bonnin P, Bazzi-Grossin C et al (1997) Evidence of fetal cerebral vasodilatation induced by submaximal maternal dynamic exercise in human pregnancy. J Perinat Med 25(1):63–70

Bornstein MH, DiPietro JA et al (2002) Prenatal cardiac function and postnatal cognitive development: an exploratory study. Infancy 3(4):475–494

Botkin C, Driscoll CE (1991) Maternal aerobic exercise: newborn effects. Fam Pract Res J 11(4):387–393

Braun T, Sierra F et al (2004) Continuous telemetric monitoring of fetal oxygen partial pressure during labor. Arch Gynecol Obstet 270(1):40–45

Brenner IK, Wolfe LA et al (1999) Physical conditioning effects on fetal heart rate responses to graded maternal exercise. Med Sci Sports Exerc 31(6):792–799

Brooten D, Youngblut JM et al (2001) A randomized trial of nurse specialist home care for women with high-risk pregnancies: outcomes and costs. Am J Manag Care 7(8):793–803

Campbell MK, Mottola MF (2001) Recreational exercise and occupational activity during pregnancy and birth weight: a case-control study. Am J Obstet Gynecol 184(3):403–408

Carey GB, Quinn TJ (1998) Effect of exercise on milk and nursing babies. Med Sci Sports Exerc 30(11):1659–1660

Carey GB, Quinn TJ et al (1997) Breast milk composition after exercise of different intensities. J Hum Lact 13(2):115–120

Carmichael SL, Shaw GM et al (2002) Physical activity and risk of neural tube defects. Matern Child Health J 6(3):151–157

Carpenter MW, Sady SP et al (1988) Fetal heart rate response to maternal exertion. J Am Med Assoc 259(20):3006–3009

Chaddha V, Simchen MJ et al (2005) Fetal response to maternal exercise in pregnancies with uteroplacental insufficiency. Am J Obstet Gynecol 193(3 Pt 2):995–999

Chasan-Taber L, Schmidt MD et al (2004) Development and validation of a pregnancy physical activity questionnaire. Med Sci Sports Exerc 36(10):1750–1760

Clapp JF (2006) Influence of endurance exercise and diet on human placental development and fetal growth. Placenta 27(6–7):527–534

Clapp JF 3rd (1985) Fetal heart rate response to running in midpregnancy and late pregnancy. Am J Obstet Gynecol 153(3):251–252

Clapp JF 3rd (1989) The effects of maternal exercise on early pregnancy outcome. Am J Obstet Gynecol 161(6 Pt 1):1453–1457

Clapp JF 3rd (1990) The course of labor after endurance exercise during pregnancy. Am J Obstet Gynecol 163(6 Pt 1):1799–1805

Clapp JF 3rd (1991) Exercise and fetal health. J Dev Physiol 15(1):9–14

Clapp JF 3rd (1994) A clinical approach to exercise during pregnancy. Clin Sports Med 13(2):443–458

Clapp JF 3rd (1996) Morphometric and neurodevelopmental outcome at age five years of the offspring of women who continued to exercise regularly throughout pregnancy. J Pediatr 129(6):856–863

Clapp JF 3rd (1998) Effect of dietary carbohydrate on the glucose and insulin response to mixed caloric intake and exercise in both nonpregnant and pregnant women. Diabetes Care 21(Suppl 2):B107–112

Clapp JF 3rd (2000) Exercise during pregnancy. A clinical update. Clin Sports Med 19(2):273–286

Clapp JF 3rd (2003) The effects of maternal exercise on fetal oxygenation and feto-placental growth. Eur J Obstet Gynecol Reprod Biol 110(Suppl 1):S80–85

Clapp JF 3rd, Capeless EL (1990) Neonatal morphometrics after endurance exercise during pregnancy. Am J Obstet Gynecol 163(6 Pt 1):1805–1811

Clapp JF 3rd, Dickstein S (1984) Endurance exercise and pregnancy outcome. Med Sci Sports Exerc 16(6):556–562

Clapp JF 3rd, Kim H et al (2000) Beginning regular exercise in early pregnancy: effect on fetoplacental growth. Am J Obstet Gynecol 183(6):1484–1488

Clapp JF 3rd, Kim H et al (2002) Continuing regular exercise during pregnancy: effect of exercise volume on fetoplacental growth. Am J Obstet Gynecol 186(1):142–147

Clapp JF 3rd, Little KD (1994) The physiological response of instructors and participants to three aerobics regimens. Med Sci Sports Exerc 26(8):1041–1046

Clapp JF 3rd, Little KD et al (1995) The effect of regular maternal exercise on erythropoietin in cord blood and amniotic fluid. Am J Obstet Gynecol 172(5):1445–1451

Clapp JF 3rd, Little KD et al (1993) Fetal heart rate response to sustained recreational exercise. Am J Obstet Gynecol 168(1 Pt 1):198–206

Clapp JF 3rd, Little KD et al (2003) Effect of maternal exercise and fetoplacental growth rate on serum erythropoietin concentrations. Am J Obstet Gynecol 188(4):1021–1025

Clapp JF 3rd, Lopez B et al (1999) Neonatal behavioral profile of the offspring of women who continued to exercise regularly throughout pregnancy. Am J Obstet Gynecol 180 (1 Pt 1):91–94

Clapp JF 3rd, Simonian S et al (1998) The one-year morphometric and neurodevelopmental outcome of the offspring of women who continued to exercise regularly throughout pregnancy. Am J Obstet Gynecol 178(3):594–599

Clapp JF, Lopez B (2007) Low-versus high-glycemic index diets in women: effects on caloric requirement, substrate utilization and insulin sensitivity. Metab Syndr Relat Disord 5(3):231–242

Clarke PE, Gross H (2004) Women's behaviour, beliefs and information sources about physical exercise in pregnancy. Midwifery 20(2):133–141

Clothier B, Stringer M et al (2007) Periodontal disease and pregnancy outcomes: exposure, risk and intervention. Best Pract Res Clin Obstet Gynaecol 21(3):451–466

Collings C, Curet LB (1985) Fetal heart rate response to maternal exercise. Am J Obstet Gynecol 151(4):498–501

Collings CA, Curet LB et al (1983) Maternal and fetal responses to a maternal aerobic exercise program. Am J Obstet Gynecol 145(6):702–707

Cooper KA, Hunyor SN et al (1987) Fetal heart rate and maternal cardiovascular and catecholamine responses to dynamic exercise. Aust N Z J Obstet Gynaecol 27(3):220–223

Curet LB, Orr JA et al (1976) Effect of exercise on cardiac output and distribution of uterine blood flow in pregnant ewes. J Appl Physiol 40(5):725–728

D'Angelo D, Williams L, Morrow B, Cox S, Harris N, Harrison L, Posner SF, Hood JR, Zapata L (2007) Preconception and interconception health status of women who recently gave birth to a live-born infant—pregnancy risk assessment monitoring system (PRAMS), United States, 26 reporting areas, 2004. Morb Mortal Wkly Rep 55(RR-6):1–35

Dale E, Mullinax KM et al (1982) Exercise during pregnancy: effects on the fetus. Can J Appl Sport Sci 7(2):98–103

David M, Hirsch M et al (2007) An estimate of fetal autonomic state by time-frequency analysis of fetal heart rate variability. J Appl Physiol 102(3):1057–1064

Davies GA, Wolfe LA et al (2003) Joint SOGC/CSEP clinical practice guideline: exercise in pregnancy and the postpartum period. Can J Appl Physiol 28(3):330–341

de Barros MC, Lopes MA et al (2010) Resistance exercise and glycemic control in women with gestational diabetes mellitus. Am J Obstet Gynecol 203(6):556 e551–556

Dewey KG, McCrory MA (1994) Effects of dieting and physical activity on pregnancy and lactation. Am J Clin Nutr 59(2 Suppl): 446S–452S (discussion 452S-453S)

DiPietro JA, Bornstein MH et al (2002a) What does fetal movement predict about behavior during the first two years of life? Dev Psychobiol 40(4):358–371

DiPietro JA, Bornstein MH et al (2007) Fetal heart rate and variability: stability and prediction to developmental outcomes in early childhood. Child Dev 78(6):1788–1798

DiPietro JA, Caulfield L et al (2004) Fetal neurobehavioral development: a tale of two cities. Dev Psychol 40(3):445–456

DiPietro JA, Costigan KA et al (2002b) Fetal state concordance predicts infant state regulation. Early Hum Dev 68(1):1–13

DiPietro JA, Hodgson DM et al (1996a) Development of fetal movement–fetal heart rate coupling from 20 weeks through term. Early Hum Dev 44(2):139–151

DiPietro JA, Hodgson DM et al (1996b) Fetal neurobehavioral development. Child Dev 67(5):2553–2567

Dipietro JA, Irizarry RA et al (2001) Cross-correlation of fetal cardiac and somatic activity as an indicator of antenatal neural development. Am J Obstet Gynecol 185(6):1421–1428

DiPietro JA, Kivlighan KT et al (2010) Prenatal antecedents of newborn neurological maturation. Child Dev 81(1):115–130

Domingues MR, Barros AJ et al (2008) Leisure time physical activity during pregnancy and preterm birth in Brazil. Int J Gynaecol Obstet 103(1):9–15

Doussard-Roosevelt JA, McClenny BD et al (2001) Neonatal cardiac vagal tone and school-age developmental outcome in very low birth weight infants. Dev Psychobiol 38(1):56–66

Duncombe D, Wertheim EH et al (2009) Factors related to exercise over the course of pregnancy including women's beliefs about the safety of exercise during pregnancy. Midwifery 25(4):430–438

Dye TD, Knox KL et al (1997) Physical activity, obesity, and diabetes in pregnancy. Am J Epidemiol 146(11):961–965

Elia M, Betts P et al (2007) Fetal programming of body dimensions and percentage body fat measured in prepubertal children with a 4-component model of body composition, dual-energy X-ray absorptiometry, deuterium dilution, densitometry, and skinfold thicknesses. Am J Clin Nutr 86(3):618–624

Evenson KR, Bradley CB (2010) Beliefs about exercise and physical activity among pregnant women. Patient Educ Couns 79(1):124–129

Evenson KR, Moos MK et al (2009) Perceived barriers to physical activity among pregnant women. Matern Child Health J 13(3):364–375

Evenson KR, Wen F (2010a) Measuring physical activity among pregnant women using a structured one-week recall questionnaire: evidence for validity and reliability. Int J Behav Nutr Phys Act 7:21

Evenson KR, Wen F (2010b) National trends in self-reported physical activity and sedentary behaviors among pregnant women: NHANES 1999–2006. Prev Med 50(3):123–128

Fell DB, Joseph KS et al (2009) The impact of pregnancy on physical activity level. Matern Child Health J 13(5):597–603

Finch CF (2002) The risk of abdominal injury to women during sport. J Sci Med Sport 5(1):46–54

Finley JP, Nugent ST (1995) Heart rate variability in infants, children and young adults. J Auton Nerv Syst 51(2):103–108

Fleten C, Stigum H et al (2010) Exercise during pregnancy, maternal prepregnancy body mass index, and birth weight. Obstet Gynecol 115(2 Pt 1):331–337

Florack EI, Pellegrino AE et al (1995) Influence of occupational physical activity on pregnancy duration and birthweight. Scand J Work Environ Health 21(3):199–207

Fox NA, Porges SW (1985) The relation between neonatal heart period patterns and developmental outcome. Child Dev 56(1):28–37

Gavard JA, Artal R (2008) Effect of exercise on pregnancy outcome. Clin Obstet Gynecol 51(2):467–480

Giroux I, Inglis SD et al (2006) Dietary intake, weight gain, and birth outcomes of physically active pregnant women: a pilot study. Appl Physiol Nutr Metab 31(5):483–489

Graca LM, Cardoso CG et al (1991) Acute effects of maternal cigarette smoking on fetal heart rate and fetal body movements felt by the mother. J Perinat Med 19(5):385–390

Hall DC, Kaufmann DA (1987) Effects of aerobic and strength conditioning on pregnancy outcomes. Am J Obstet Gynecol 157(5):1199–1203

Halmesmaki E, Ylikorkala O (1986) The effect of maternal ethanol intoxication on fetal cardiotocography: a report of four cases. Br J Obstet Gynaecol 93(3):203–205

Hatch MC, Shu XO et al (1993) Maternal exercise during pregnancy, physical fitness, and fetal growth. Am J Epidemiol 137(10):1105–1114

Hatoum N, Clapp JF 3rd et al (1997) Effects of maternal exercise on fetal activity in late gestation. J Matern Fetal Med 6(3):134–139

Hauth JC, Gilstrap LC 3rd et al (1982) Fetal heart rate reactivity before and after maternal jogging during the third trimester. Am J Obstet Gynecol 142(5):545–547

Higgins M, Galvin D et al (2011) Pregnancy in women with Type 1 and Type 2 diabetes in Dublin. Ir J Med Sci 180(2):469–473

Hopkins SA, Baldi JC et al (2011) Effects of exercise training on maternal hormonal changes in pregnancy. Clin Endocrinol 74(4):495–500

Horns PN, Ratcliffe LP et al (1996) Pregnancy outcomes among active and sedentary primiparous women. J Obstet Gynecol Neonatal Nurs 25(1):49–54

Houghton PE, Mottola MF et al (2000) Effect of maternal exercise on fetal and placental glycogen storage in the mature rat. Can J Appl Physiol 25(6):443–452

Jarrett JC 2nd, Spellacy WN (1983) Jogging during pregnancy: an improved outcome? Obstet Gynecol 61(6):705–709

Joshi PP (1999) Patient compliance with drug therapy. J Assoc Physicians India 47(6):655–656

Jovanovic L, Kessler A et al (1985) Human maternal and fetal response to graded exercise. J Appl Physiol 58(5):1719–1722

Juhl M, Madsen M, Andersen AM, Andersen PK, Olsen J (2012) Distribution and predictors of exercise habits among pregnant women in the Danish National Birth Cohort. Scand J Med Sci Sports 22(1):128–138

Juhl M, Olsen J et al (2010) Physical exercise during pregnancy and fetal growth measures: a study within the Danish National Birth Cohort. Am J Obstet Gynecol 202(1):63 e61–63 e68

Kardel KR, Kase T (1998) Training in pregnant women: effects on fetal development and birth. Am J Obstet Gynecol 178(2):280–286

Kaufman C, Berg K et al (2006) Ratings of perceived exertion of ACSM exercise guidelines in individuals varying in aerobic fitness. Res Q Exerc Sport 77(1):122–130

Kennelly MM, Geary M et al (2002a) Exercise-related changes in umbilical and uterine artery waveforms as assessed by Doppler ultrasound scans. Am J Obstet Gynecol 187(3):661–666

Kennelly MM, McCaffrey N et al (2002b) Fetal heart rate response to strenuous maternal exercise: not a predictor of fetal distress. Am J Obstet Gynecol 187(3):811–816

Klebanoff MA, Shiono PH et al (1990) The effect of physical activity during pregnancy on preterm delivery and birth weight. Am J Obstet Gynecol 163(5 Pt 1):1450–1456

Koemeester AP, Broersen JP et al (1995) Physical work load and gestational age at delivery. Occup Environ Med 52(5):313–315

Koizumi K, Terui N et al (1982) Functional significance of coactivation of vagal and sympathetic cardiac nerves. Proc Natl Acad Sci U S A 79(6):2116–2120

Kramer MS (2000) Regular aerobic exercise during pregnancy. Cochrane database of systematic reviews(2):CD000180

Krans EE, Gearhart JG et al (2005) Pregnant women's beliefs and influences regarding exercise during pregnancy. J Miss State Med Assoc 46(3):67–73

LaPorte RE, Montoye HJ et al (1985) Assessment of physical activity in epidemiologic research: problems and prospects. Public Health Rep 100(2):131–146

Leiferman JA, Evenson KR (2003) The effect of regular leisure physical activity on birth outcomes. Matern Child Health J 7(1):59–64

Lindholm ES, Norman M et al (2010) Weight control program for obese pregnant women. Acta Obstet Gynecol Scand 89(6):840–843

Lovelady CA, Lonnerdal B et al (1990) Lactation performance of exercising women. Am J Clin Nutr 52(1):103–109

MacPhail A, Davies GA et al (2000) Maximal exercise testing in late gestation: fetal responses. Obstet Gynecol 96(4):565–570

Madsen M, Jorgensen T et al (2007) Leisure time physical exercise during pregnancy and the risk of miscarriage: a study within the Danish National Birth Cohort. Bjog 114(11):1419–1426

Magann EF, Evans SF et al (2002) Antepartum, intrapartum, and neonatal significance of exercise on healthy low-risk pregnant working women. Obstet Gynecol 99(3):466–472

Malnick SD, Knobler H (2006) The medical complications of obesity. QJM 99(9):565–579

Manders MA, Sonder GJ et al (1997) The effects of maternal exercise on fetal heart rate and movement patterns. Early Hum Dev 48(3):237–247

Marquez-Sterling S, Perry AC et al (2000) Physical and psychological changes with vigorous exercise in sedentary primigravidae. Med Sci Sports Exerc 32(1):58–62

Marsal K, Lofgren O et al (1979) Fetal breathing movements and maternal exercise. Acta Obstet Gynecol Scand 58(2):197–201

May L, Glaros AG, Yeh H-W, Clapp JF, Gustafson KM (2010) Aerobic exercise during pregnancy influences fetal cardiac autonomic control of heart rate and heart rate variability. Early Hum Dev 86(4):17

May LE, Gustafson KF et al (2008) Effects of maternal exercise on the fetal heart. FASEB J 22(1_MeetingAbstracts): 1175.1173

May LE, Gustafson KM (2009) Differences among exercisers and non-exercisers during pregnancy. FASEB J 23(1_MeetingAbstracts): 955.911

May L, Suminski RS, Langaker M, Yeh H-W, Gustafson KM (2011a) Regular Maternal Exercise Dose and Fetal Heart Outcome. Medicine and Science in Sports and Exercise (in print)

May LE, Suminski RR, Yeh H-W, Gustafson KM (2011b) Maternal physical activity dose and infant heart adaptations. J Dev Origins Health Dis (Meeting abstract)

McMurray RG, Katz VL et al (1995) Maternal and fetal responses to low-impact aerobic dance. Am J Perinatol 12(4):282–285

Misra DP, Strobino DM, Stashinko EE, Nagey DA, Nanda J (1998) Effects of physical activity on preterm birth. American journal of epidemiology 147(7):628–635

Misra R, Bhowmik D et al (2003) Pregnancy with chronic kidney disease: outcome in Indian women. J Womens Health (Larchmt) 12(10):1019–1025

Mochizuki M, Maruo T et al (1985) Mechanism of foetal growth retardation caused by smoking during pregnancy. Acta Physiol Hung 65(3):295–304

Mottola MF (2009) Exercise prescription for overweight and obese women: pregnancy and postpartum. Obstet Gynecol Clin North Am 36(2): 301–316, viii

Mottola MF, Bagnall KM et al (1989) Effects of strenuous maternal exercise on fetal organ weights and skeletal muscle development in rats. J Dev Physiol 11(2):111–115

Mudd LM, Nechuta S et al (2009) Factors associated with women's perceptions of physical activity safety during pregnancy. Prev Med 49(2–3):194–199

Murray M, Huelsmann G (2009) Labor and delivery nursing : a guide to evidence-based practice. Springer, New York

Nagai A, Sakamoto K et al (1993) The effect of maternal exercise on somatic growth and lung development of fetal rats: morphologic and morphometric studies. Pediatr Pulmonol 15(6):332–338

Nagaoka S, Bito Y et al (2003) Comparative physiology of postnatal developments of cardiopulmonary reflex. Uchu Seibutsu Kagaku 17(3):265–266

Nijhuis IJ, ten Hof J (1999) Development of fetal heart rate and behavior: indirect measures to assess the fetal nervous system. Eur J Obstet Gynecol Reprod Biol 87(1):1–2

Nijhuis JG, Staisch KJ et al (1984) A sinusoidal-like fetal heart-rate pattern in association with fetal sucking–report of 2 cases. Eur J Obstet Gynecol Reprod Biol 16(5):353–358

O'Connor PJ, Poudevigne MS et al (2011) Safety and efficacy of supervised strength training adopted in pregnancy. J Phys Act Health 8(3):309–320

O'Neill ME (1996) Maternal rectal temperature and fetal heart rate responses to upright cycling in late pregnancy. Br J Sports Med 30(1):32–35

Oostdam N, van Poppel MN et al (2009) Design of FitFor2 study: the effects of an exercise program on insulin sensitivity and plasma glucose levels in pregnant women at high risk for gestational diabetes. BMC pregnancy and childbirth 9:1

Oppenheimer LW, Lewinsky RM (1994) Power spectral analysis of fetal heart rate. Baillieres Clin Obstet Gynaecol 8(3):643–661

Paolone AM, Shangold M et al (1987) Fetal heart rate measurement during maternal exercise–avoidance of artifact. Med Sci Sports Exerc 19(6):605–609

Paolone AM, Shangold MM (1987) Artifact in the recording of fetal heart rates during material exercise. J Appl Physiol 62(2):848–849

Parizkova J (1975) Impact of daily work-load during pregnancy on the microstructure of the rat heart in male offspring. Eur J Appl Physiol Occup Physiol 34(4):323–326

Parizkova J (1979) Cardiac microstructure in female and male offspring of exercised rat mothers. Acta Anat (Basel) 104(4):382–387

Petersen AM, Leet TL et al (2005) Correlates of physical activity among pregnant women in the United States. Med Sci Sports Exerc 37(10):1748–1753

Phillips JB, Billson VR et al (2009) Autopsy standards for fetal lengths and organ weights of an Australian perinatal population. Pathology 41(6):515–526

Pivarnik JM, Chambliss HO, Clapp JF, Dugan SA, Hatch MC, Lovelady CA, Mottola MF, Williams MA (2006) Impact of physical activity during pregnancy and postpartum on chronic disease risk. Med Sci Sports Exerc 38(5):989–1006

Platt LD, Artal R et al (1983) Exercise in pregnancy II. Fetal responses. Am J Obstet Gynecol 147(5):487–491

Pomerance JJ, Gluck L et al (1974) Physical fitness in pregnancy: its effect on pregnancy outcome. Am J Obstet Gynecol 119(7):867–876

Porges SW (1972) Heart rate variability and deceleration as indexes of reaction time. J Exp Psychol 92(1):103–110

Porges SW, Humphrey MM (1977) Cardiac and respiratory responses during visual search in nonretarded children and retarded adolescents. Am J Ment Retard 82(2):162–169

Prentice A (1994) Should lactating women exercise? Nutr Rev 52(10):358–360

Quinn TJ, Carey GB (1999) Does exercise intensity or diet influence lactic acid accumulation in breast milk? Med Sci Sports Exerc 31(1):105–110

Rabkin CS, Anderson HR et al (1990) Maternal activity and birth weight: a prospective, population-based study. Am J Epidemiol 131(3):522–531

Radnai M, Pal A et al (2009) Benefits of periodontal therapy when preterm birth threatens. J Dent Res 88(3):280–284

Rao VG, Sugunan AP et al (1998) Nutritional deficiency disorders and high mortality among children of the Great Andamanese tribe. Natl Med J India 11(2):65–68

Renou P, Newman W et al (1969) Autonomic control of fetal heart rate. Am J Obstet Gynecol 105(6):949–953

Rice PL, Fort IL (1991) The relationship of maternal exercise on labor, delivery and health of the newborn. J Sports Med Phys Fitness 31(1):95–99

Rose NC, Haddow JE et al (1991) Self-rated physical activity level during the second trimester and pregnancy outcome. Obstet Gynecol 78(6):1078–1080

Schramm WF, Stockbauer JW et al (1996) Exercise, employment, other daily activities, and adverse pregnancy outcomes. Am J Epidemiol 143(3):211–218

Sharieff GQ, Rao SO (2006) The pediatric ECG. Emerg Med Clin North Am 24(1):195–208, vii–viii

Siega-Riz AM, Laraia B (2006) The implications of maternal overweight and obesity on the course of pregnancy and birth outcomes. Matern Child Health J 10(5 Suppl):S153–156

Smith E, Zare-Maivan E, May LE (2008) 52nd Annual AOA research conference–abstracts. J Am Osteopath Assoc 108(8):413–454

Sontag LW, Huff E (1938) A timer for use with a Westinghouse moving coil oscillograph. Science 88(2289):459–460

Spinnewijn WE, Lotgering FK et al (1996a) Fetal heart rate and uterine contractility during maternal exercise at term. Am J Obstet Gynecol 174(1 Pt 1):43–48

Spinnewijn WE, Wallenburg HC et al (1996b) Peak ventilatory responses during cycling and swimming in pregnant and nonpregnant women. J Appl Physiol 81(2):738–742

Steegers EA, Buunk G et al (1988) The influence of maternal exercise on the uteroplacental vascular bed resistance and the fetal heart rate during normal pregnancy. Eur J Obstet Gynecol Reprod Biol 27(1):21–26

Sternfeld B, Quesenberry CP Jr et al (1995) Exercise during pregnancy and pregnancy outcome. Med Sci Sports Exerc 27(5):634–640

Stevenson ET, Davy KP et al (1997) Blood pressure risks factors in healthy postmenopausal women: physical activity and hormone replacement. J Appl Physiol 82(2):652–660

Ten Hof J, Nijhuis IJ et al (2002) Longitudinal study of fetal body movements: nomograms, intrafetal consistency, and relationship with episodes of heart rate patterns a and b. Pediatr Res 52(4):568–575

van Doorn MB, Lotgering FK et al (1992) Maternal and fetal cardiovascular responses to strenuous bicycle exercise. Am J Obstet Gynecol 166(3):854–859

Van Leeuwen P, Geue D et al (2003) Changes in the frequency power spectrum of fetal heart rate in the course of pregnancy. Prenat Diagn 23(11):909–916

Vanderlei LC, Pastre CM et al (2009) Basic notions of heart rate variability and its clinical applicability. Revista brasileira de cirurgia cardiovascular : orgao oficial da Sociedade Brasileira de Cirurgia Cardiovascular 24(2):205–217

Wallace JP, Ernsthausen K et al (1992a) The influence of the fullness of milk in the breasts on the concentration of lactic acid in postexercise breast milk. Int J Sports Med 13(5):395–398

Wallace JP, Inbar G et al (1992b) Infant acceptance of postexercise breast milk. Pediatrics 89(6 Pt 2):1245–1247

Watson WJ, Katz VL et al (1991) Fetal responses to maximal swimming and cycling exercise during pregnancy. Obstet Gynecol 77(3):382–386

Webb KA, Wolfe LA et al (1994) Effects of acute and chronic maternal exercise on fetal heart rate. J Appl Physiol 77(5):2207–2213

Weir Z, Bush J et al (2010) Physical activity in pregnancy: a qualitative study of the beliefs of overweight and obese pregnant women. BMC pregnancy and childbirth 10:18

Winn HN, Hess O et al (1994) Fetal responses to maternal exercise: effect on fetal breathing and body movement. Am J Perinatol 11(4):263–266